Smart Specialization in Croatia

Smart Specialization in Croatia

Inputs from Trade, Innovation, and Productivity Analysis

Arabela Aprahamian and Paulo Guilherme Correa, Editors

WORLD BANK GROUP

CIIP Competitive Industries and Innovation Program

Financed by

in partnership with
WORLD BANK GROUP

Contents

Map

Tables

Acknowledgments

This report was developed by a team led by Arabela Aprahamian, senior operations officer, and Paulo Correa, lead economist, and comprised of Mariana Iootty, Daria Taglioni, Ana Paula Cusolito, Gabriela Elizondo Azuela, Michael Ferrantino, Swarmim Wagle, and Ana Florina Pirlea. The team is grateful for the input provided by Deborah Winkler, Guillermo Arenas, Claire Hollweg, Jose Daniel Reyes, Sanja Zivkovic, Asad Ali Ahmed, and Javier Inon (all from the World Bank) and by Sonja Radas, Bruno Škrinjarić, Jurica Jug-Dujakovic, Ana Gavrilovic, Miroslav Božić, and Danica Ramljak (external experts). The report benefited from the invaluable advice, comments, and suggestions of the peer reviewers, including Theo David Thomas, lead economist; Donato De Rosa, senior economist; and Ivan Rossignol, chief technical specialist (World Bank). Financial support from the Competitive Industries and Innovation Program (CIIP) is gratefully acknowledged.

Abbreviations

BERD	Business enterprise R&D
BICRO	Business Innovation Croatian Agency
CIIP	Competitive Industries and Innovation Program
CRANE	Croatian business angel network
EC	European Commission
ECA	Europe and Central Asia
EU	European Union
FDI	Foreign direct investment
FINA	Croatian Financial Agency
GDP	Gross domestic product
GVC	Global value chain
HRV	Croatia
HS	Harmonized System
ICT	Information and communications technology
KI	Knowledge-intensive
MEPPC	Ministry of Environmental Protection, Physical Planning, and Construction
MSES	Ministry of Science, Education, and Sports
NAFTA	North American Free Trade Agreement
NBE	New big-entries
NSU	New start-ups
NUTS	Nomenclature of Territorial Units for Statistics
OECD	Organisation for Economic Co-operation and Development
R&D	Research and development
RCA	Revealed comparative advantage
RIS3	Research and innovation strategy for smart specialization
RQE	Relative quality elasticity
SITC	Standard International Trade Classification
SMEs	Small and medium enterprises

SOE	State-owned enterprise
TECHCRO	Technology Infrastructure Development Programme
TFP	Total factor productivity
UKF	Unity through Knowledge Fund
WBGES	World Bank Group Entrepreneurship Snapshots

Executive Summary

Missing Out on the Benefits of Research and Innovation

For Croatia—a small country of some 4.3 million—exports are critical for growth and employment generation. But its export openness is lower than in countries with similar incomes. Its per capita gross domestic product (GDP) is stagnant. And its trade and development performance has been lackluster since the financial crisis, especially against such peer countries as Estonia and Slovenia, which have sharply expanded their shares of exports in GDP.

Low export diversification, weak competitiveness, and little technological sophistication explain the stagnant exports. In 2012, five sectors made up about two-thirds of the export basket; and of 3,400 products at HS-6 level, only four were "winners" in growing sectors. Croatia has been specializing in its existing markets, but its expansion into new products and markets has been limited. And for many exports the complexity and sophistication have been low to medium.

The fundamental problem is the failure to renew and transform the manufacturing base, linked to low rates of firm entry and exit. Annual entry rates are only 5.5 percent, far lower than the 9–18 percent for peers. Annual exit rates are 6.5 percent, against 7–26 percent for peers. In the case of some comparators, firm entry largely outpaces firm exit—marking the transition to a market economy. But for Croatia, exits outpace entries, reinforcing the view that the country has a stagnant economy with little creative destruction or innovation—hence the limited export diversification.

Another marker of economic stagnation is the inadequate levels of R&D by Croatian enterprises. The country's business enterprise R&D (BERD) as a share of GDP is lower than the EU-27 average—and is falling farther behind. Its additional financing needs to reach the Europe 2020 Agenda target for R&D spending: 0.65 percent of GDP—one-third government and two-thirds private.

Weak governance is holding back the impact of R&D spending. Public funds are allocated without clear priorities or a results orientation, reflected in low government support to businesses and low shares of experimental (as opposed to basic and applied) research in the total. An opaque governance structure spawns a policy-making and implementation system that lacks cohesion and that suits the needs of individual agencies rather than the overall system. Similarly, technology and innovation policy is fragmented, leading to programs with overlapping objectives and wasting resources. For example, the Ministry of Science, Education, and Sports (MSES)—responsible for planning, funding, and monitoring the overall science and education system—and the Ministry of Economy and the Ministry of Entrepreneurship—with programs to increase business–industry linkages—do not fully coordinate their activities.

Policy Implications

Croatia's research and innovation strategy for smart specialization (RIS3) is a good opportunity for the country to hoist itself from its low-level equilibrium. RIS3 can promote conditions for a more vigorous structural transformation and a reinvigorated productive sector, through the following measures:

Addressing weakness in the business environment and promoting entry and exit reforms to spur productivity. Croatia should reduce sector-specific state aid to minimize distortions in the economy and to facilitate competition. Given the key role of services as intermediate inputs for many enterprises, the authorities should ensure the full implementation of the Services Directive of the European Commission and remove the remaining barriers to foreign direct investment in services. Improving insolvency procedures—by reducing the time for debt recovery and increasing the recovery rate—and reducing the time needed to enforce a contract would also increase firm productivity.

Supporting innovation through R&D investments in small and medium enterprises (SMEs) and improving the research system's performance. Croatia needs to increase, gradually but substantially, direct support of business investments in R&D. It can do this through the Proof of Concept and RAZUM programs of the Business Innovation Center.[1] The authorities can also consider developing a new program, possibly financed by European Structural and Investment Funds, to support angel and venture capital services. The focus of these funds in innovation should be to promote young firms' and SMEs' expenditures on R&D. Support for enterprises is foreseen in the Croatia 2014–2020 Operational Program for Competitiveness and Cohesion. The focus is on acquiring new manufacturing technologies, equipment, and machinery; underpinning technology transfers from scientific and research organizations; and encouraging the use of key technologies for developing new products, services, and business models.

To improve research excellence, Croatia can streamline the innovation framework by strengthening the connection with the global scientific community (for example, improving conditions for mobility of researchers). It can also reform the research profession to emphasize excellence in career development. And it can promote more use of the research infrastructure through an open-access policy.

Several strategic investments in the country's innovation system are planned as part of the Operational Program:

- Bolstering R&D infrastructure and equipment in public and private research institutes and organizations.
- Building, furnishing, and covering the initial operating costs of private and public institutions supporting the commercialization of innovation and technology transfer.
- Establishing highly focused centers of competencies.
- Establishing a high-technology network to improve the research environment by ensuring access to scientific databases, scientific publications and journals, and digital resources.
- Developing professional services for knowledge and technology transfers.
- Raising awareness of the benefits of intellectual property protection and technology transfer in public academic and research institutes.

Strengthening the governance of innovation policy may be the biggest challenge for boosting research and innovation impact. This is true both for managing European Structural Funds and Strategic Investments and for implementing the policy reforms to support the programs that these funds finance. In designing and implementing RIS3, the authorities should use a results-based approach combined with a fully integrated monitoring and evaluation system. This approach would allow for structured learning and systematic adjustment of policies and programs in the pursuit of predefined objectives. It is essential to replace the emphasis on defining sectors and committing resources up front with a results-based approach that allows some flexibility for policy and program experimentation and allocations of resources based on results.

Croatia could also consider:

- Strengthening policy-making and coordination bodies at the level of prime minister, with input from the private sector.
- Creating a technical secretariat and adopting a results-driven monitoring and evaluation framework for public expenditures on R&D and innovation.
- Strengthening the BICRO-managed programs implemented by HAMAG-BICRO as the agency in charge of promoting R&D investments in SMEs and corresponding programs.
- Accelerating adoption of the Research Infrastructure Roadmap-an EU requirement—and of open access to resources by the scientific community.
- Improving a merit-driven selection process, subject to international peer review and transparency.
- Promoting proper impact assessments, public consultations, and systematic reviews of policies and programs.
- Speeding implementation of the Western Balkans Regional R&D Strategy for Innovation.

Fully integrating monitoring and evaluation with policies and programs is essential to allowing systematic learning and improvement. Rather than ex ante targeting of certain sectors, research and innovation policies should identify intermediate goals and the market and institutional failures that reduce chances of success. Public actions would then tentatively aim at correcting such failures. They would be monitored, evaluated, and adjusted in response to lessons acquired through systematic learning. That there are no predefined policy recipes when economic specialization is not immediately evident implies the need to raise the relevance of results orientation and to integrate proper monitoring and evaluation mechanisms with policy and program implementation.

Also essential is to focus on policy design and implementation in order to promote sectors with apparent or latent comparative advantage. Targeting is more likely to achieve its goals if Croatia's comparative advantages are known to policy makers who can then gear economic policies to exploit those advantages. But if information about specialization is unavailable, the right policy mix is more likely to involve interventions that enable self-discovery and that avoid the trap of picking winners in an environment with scarce information. The nature of entrepreneurial activity is to experiment with new product niches, discover the cost, and abandon those that cannot be undertaken at low enough cost to be profitable.

Note

1. Currently, HAMAG-BICRO.

CHAPTER 1

Introduction

Background

In July 2013, Croatia became the 28th member of the European Union (EU). This achievement was underpinned by a decade of solid macroeconomic and institutional reforms that was marred, however, by the global financial crisis, which has exposed Croatia's structural vulnerabilities (World Bank 2014).

Croatia's economic growth until the late 2000s was impressive. In 2003–08, economic and income growth were fueled by domestic consumption, a growing current account deficit, and increasing dependence on international finance. High and sustained growth, alongside a population in decline, raised gross domestic product (GDP) per capita, enabling it to start to converge—and quickly, partly because of the catch-up effect after a dramatic drop in the early 1990s—with the average of the 15 countries of the EU (EU-15; figure 1.1). More broadly, Croatia substantially improved its macroeconomic framework and kept its social indicators among the highest in Eastern Europe.

Such growth rates are unlikely to resume naturally in the near future—generating important social and economic costs. Since the start of the global crisis in 2008, the Croatian economy has been contracting steadily. Real GDP fell by 2 percent in 2012, by 0.9 percent in 2013 and estimates for 2014 point to a further decline of 0.8 percent. The cumulative decrease over 2008–13 is estimated at 12 percent. In the EU, Croatia and Greece have had the longest recessions. Croatia's recovery will be slow, such that a simple linear extrapolation of recent average growth suggests that the country's per capita income in even 50 years will hit only 60 percent of that in the United States, a level reached by EU-27 countries in 2000. Croatia could reach this level in less than one-third of the time by raising its annual growth rate by some 1–1.5 percentage points and by sustaining such rates for little more than a decade. If it fails to do that, the forgone opportunity of higher living standards is a cost to be borne mainly by future generations.

As highlighted by a World Bank–EU report (World Bank 2009), to accelerate growth in the coming decades the country needs to shift toward a more productivity-based and export-led growth pattern. To do this, policy makers have

Figure 1.1 Croatian GDP per Capita and Growth Rates

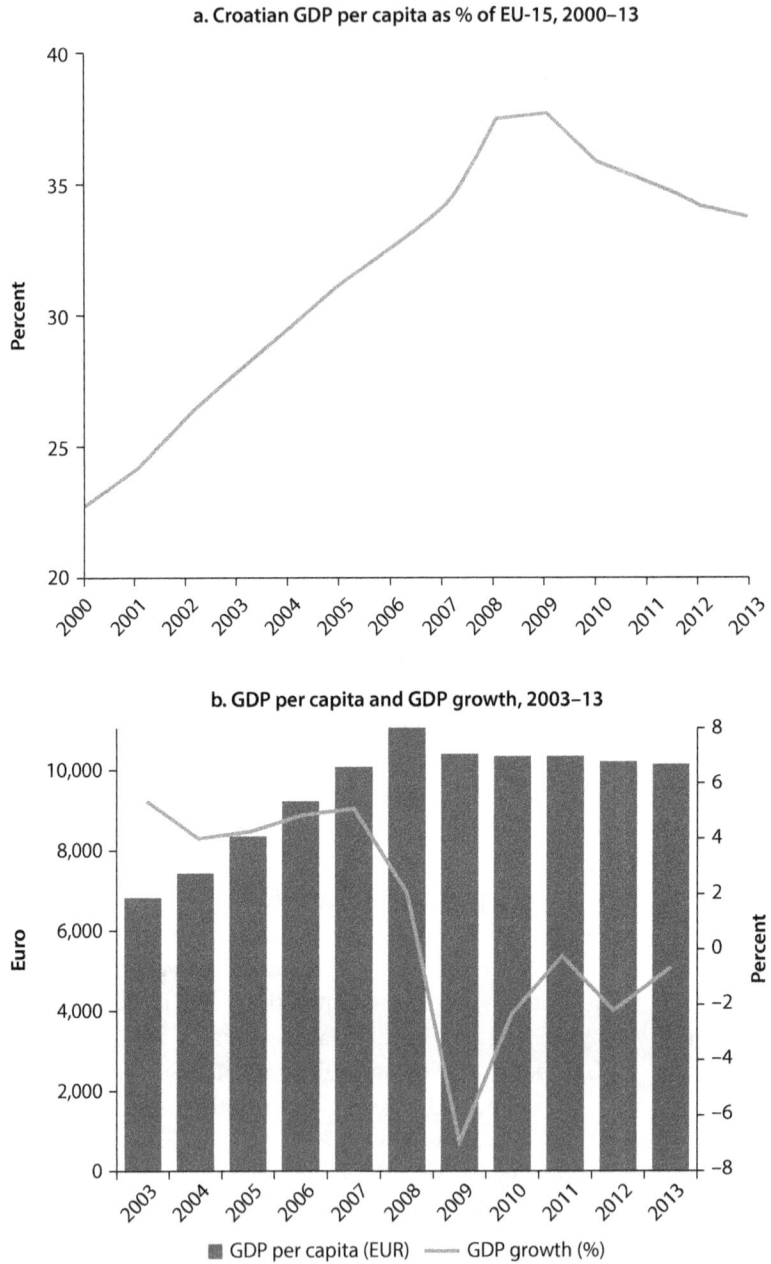

a. Croatian GDP per capita as % of EU-15, 2000–13

b. GDP per capita and GDP growth, 2003–13

■ GDP per capita (EUR) ——— GDP growth (%)

Sources: Eurostat; Elaborations based on data from the Croatian Bureau of Statistics: *Statistical Yearbook 2013, Statistical Yearbook 2010.*

to resolve four main challenges: increasing the employment level, raising productivity, fostering innovation, and deepening trade integration. With high investment ratios (and an apparently small elasticity of growth in response to increases in investment compared with the fast-growing economies in the region), a strategy of expanding potential output based primarily on further capital accumulation does not seem realistic. Further, Croatia's demographic trends (aging population, low birth rates, and imminent decreases in working-age cohorts) will limit the contribution of labor to the expansion of economic output over the longer term. Policies focused on raising labor participation and reducing unemployment can, however, help increase the contribution of labor to economic growth, as can policies to raise aggregate productivity by addressing the factors that hinder market dynamism. These measures, beyond the innovation-enabling policies discussed below, should support a focus on deepening trade integration, which would ameliorate the widening savings–investment gap, relax the economy's external financial constraints, and enable it to benefit from world demand—all improving long-term growth prospects.[1]

Deepening trade integration will require the country to further develop its comparative advantage in innovation-driven and more skill- and knowledge-intensive sectors—a core theme in its research and innovation strategy for smart specialization (RIS3). Expanding and diversifying exports is closely related to increasing productivity and fostering innovation. In studies for many countries, Wagner (2007) confirmed that more productive firms are more likely to export; equally, exporters tend to be more productive than nonexporters (Bernard and Jensen 1995, 1999, 2004a, 2004b). Robust, positive relationships between firm-level innovation and productivity, as well as some innovation inputs such as R&D and productivity, have been documented by Hall, Mairesse, and Mohnen (2010) and Hall (2011). Product innovations such as *upgrading* the quality of existing goods or services allow companies to pull ahead of the competition by differentiating their products (Cusolito 2009; Bloom, Draca, and van Reenen 2012); product innovations like the *creation* of new goods help firms to diversify their portfolios, conquer new markets, and reduce the effects of external shocks (Lederman and Klinger 2006).

Aim and Structure of the Report

Against this backdrop, this report assesses the recent trends in Croatia's performance in trade, productivity, and innovation, in order to identify priorities for its RIS3. RIS3 is an ex ante conditionality for a country to access the EU's Structural and Investment Funds in 2014–20—for which Croatia is now eligible as a new member. RIS3 can be understood on a general level as a development strategy, and as such it should identify goals and instruments, as well as combine policy reforms with strategic investments—all to promote export expansion and diversification, economic growth, and higher living standards.

This report provides selected inputs for preparing Croatia's Smart Specialization strategy. It complements the analysis of Croatia's National

Figure 1.2 Structure of the Report's Analysis

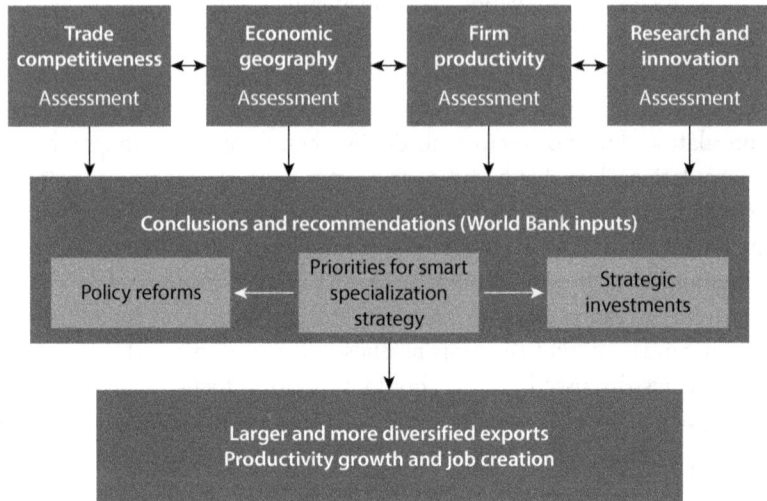

Innovation System conducted by OECD (2014) and several background studies for the country's RIS3 conducted by ECORYS (2014). The report, prepared between November 2013 and June 2014, responds to a request made by the government of Croatia and was partly funded by a grant from the Competitive Industries and Innovation Program (CIIP).

The report is based on three interrelated issues particularly relevant for developing RIS3: trade, productivity, and research and innovation. These issues were separately addressed in three background notes (which formed the basis of chapters 3, 4, and 5). The report also discusses how smart specialization can be understood (chapter 2), and it presents four case studies to illustrate possible RIS3 approaches where the level of market information differs (appendix A). The case studies present these activities not as priority sectors but as examples of potential areas for research and innovation specialization at the regional (subnational or local) level (figure 1.2).

The report's ultimate objective is to help design a strategy for a more diversified export structure, productivity growth, and job creation. It provides information on the main constraints and opportunities that Croatia faces and on the role of RIS3 in fostering economic progress.

Note

1. These challenges, including financial constraints related to high external liabilities, uncompetitive exports, as well as a corporate debt overhang and public sector indebtedness, are discussed in a recent European Commission analysis (EC Occasional Paper 179, March 2014). That paper also highlights structural weaknesses such as the poor business environment and malfunctioning labor market.

Bibliography

Bernard, A. B., and J. B. Jensen. 1995. "Exporters, Jobs, and Wages in U.S. Manufacturing: 1976–1987." *Brookings Papers on Economic Activity: Microeconomics* 1: 67–119.

———. 1999. "Exceptional Exporter Performance: Cause, Effect, or Both?" *Journal of International Economics* 47 (1): 1–25.

———. 2004a. "Exporting and Productivity in the USA." *Oxford Review of Economic Policy* 20 (3): 343–57.

———. 2004b. "Why Some Firms Export." *Review of Economics and Statistics* 86 (2): 561–69.

Bloom, N., M. Draca, and J. van Reenen. 2012. "Trade Induced Technical Change: The Impact of Chinese Imports on Innovation, Diffusion and Productivity." Working Paper, Stanford University, Palo Alto, CA.

Croatian Bureau of Statistics. 2010. *Statistical Yearbook of the Republic of Croatia 2010.* Zagreb: Croatian Bureau of Statistics. http://www.dzs.hr/default_e.htm.

———. 2013. *Statistical Yearbook 2013. Statistical Yearbook of the Republic of Croatia 2013.* Zagreb: Croatian Bureau of Statistics. http://www.dzs.hr/default_e.htm.

Cusolito, A. 2009. "Competition, Imitation, and Technical Change: Quality vs. Variety." Policy Research Working Paper 4997, World Bank, Washington, DC.

ECORYS. 2014. "Smart Specialization Strategy Croatia."

European Commission. 2014. "Macroeconomic Imbalances: Croatia 2014." Occasional Papers 179, Directorate-General for Economic and Financial Affairs, Brussels.

Hall, Bronwyn H. 2011. "Innovation and Productivity." NBER Working Paper 17178, Revised, August 2011, JEL O30.

Hall, Bronwyn H., J. Mairesse, and P. Mohnen. 2010. "Measuring the Returns to R&D." CIRANO Working Papers 2010s-02.

Lederman, D., and B. Klinger. 2006. "Diversification, Innovation, and Imitation Inside the Global Technological Frontier." Policy Research Working Paper 3872, World Bank, Washington, DC.

OECD. 2014. *OECD Reviews of Innovation Policy: Croatia 2013.* Paris: OECD Publishing. http://dx.doi.org/10.1787/9789264204362-en.

Wagner, J. 2007. "Exports and Productivity: A Survey of the Evidence from Firm-Level Data." *World Economy* 30 (1): 60–82.

World Bank. 2009. *Croatia's EU Convergence Report: Reaching and Sustaining Higher Rates of Economic Growth.* Report 48879-HR, Washington, DC: World Bank.

———. 2014. *Croatia Public Finance Review: Restructuring Spending for Stability and Growth.* Washington, DC: World Bank.

CHAPTER 2

Smart Specialization

The Guide on Research and Innovation Strategies for Smart Specialisation (the RIS3 Guide) defines national or regional research and innovation strategies for smart specialization as follows:

> integrated, place-based economic transformation agendas that … focus policy support and investments on key national/regional priorities, challenges and needs for knowledge-based development, including information and communications technology (ICT)-related measures; build on each country's/region's strengths, competitive advantages and potential for excellence; support technological as well as practice-based innovation and aim to stimulate private sector investment; get stakeholders fully involved and encourage innovation and experimentation; are evidence-based and include sound monitoring and evaluation systems. (*RIS3 Guide*, p. 8)[1]

Smart Specialization: The Main Concepts[2]

Smart specialization: in a nutshell, RIS3 may be understood as a knowledge-driven growth strategy. The idea was put forward as an instrument in the Europe 2020 Agenda.[3] It builds on the concepts developed by Foray and van Ark (2007) and David, Foray, and Hall (2009). Roughly, RIS3 can be understood as a development strategy that, building on existing comparative advantages, promotes a larger contribution of the knowledge factor to economic growth. At the enterprise level, the knowledge used by firms translates into better processes and products, new business models, or innovations; raises productivity and exports; and—in some cases—increases the number of jobs and their quality.

RIS3 may be defined as an integrated set of measures (policies, programs, and reforms) aiming to increase the impact of research and innovation on economic growth. By focusing on certain sectors, policy makers can avoid the dispersion of R&D funding (a real issue in Croatia, as will be seen later) and increase the probability of matching research results to market demand—which is especially important with, for example, the budget of €6.4 billion made available for research and innovation in 2010 by the European Commission (its largest ever), which urged member states to strengthen their policies ("Innovation Union": COM(2010)546).[4]

Regional development and economic transformation: RIS3 is expected to be applicable at the regional (i.e., local) level, as the cornerstone of regional development strategies. The objective is to trigger structural transformation in the regional economy to generate a "cluster of firms" with enough spillover effects and agglomeration economies to transform the region from "periphery" to "center." According to the RIS3 Guide (European Commission 2012), R&D, innovation, and technology policy should target certain activities with the potential to generate clusters of firms and thus a transformational effect, rather than simply promoting "scattered" innovation. Sector prioritization then becomes a core element of RIS3.

Knowledge and other productive assets: David, Foray, and Hall (2009) argued that targeted activities should stem from the entrepreneurial self-discovery process, as defined by Hausmann and Rodrik (2003). This is the decision process where entrepreneurs "discover" the markets in which to operate from a set of "modern sectors." This concept specifically rules out the use of top-down foresight exercises and similar instruments in selecting targeted sectors, emphasizing instead a bottom-up approach, including the following elements.

Defining a priority area: Smart specialization aims to target a few priority areas that are neither broad sectors nor a single firm, but that instead are new activities with high market potential. Proponents of the concept argue that this is one of its distinctive features, setting if off from traditional industrial policy.

Reconciling sector growth policy with regional development and research and innovation policies: Designing and implementing smart specialization at the regional level is challenging because of the strong agglomeration forces that arise from fundamental differences between the core and the periphery (Krugman 1991). Regional boundaries are much more open to accommodating these forces than are national boundaries, and this constitutes a challenge for the implementation of RIS3 (McCann and Ortega-Argilés 2011).

Relocating entrepreneurs: With both a diversified core and a specialized periphery, new firms may find it more profitable to relocate to cities during the process of learning and searching for their ideal domain (Duranton and Puga 2004). Although costly, this may allow them both to easily employ the different kinds of inputs needed when an experimented domain fails and to try a new one. Such a relocation may, however, curb the emergence of new activities in the periphery.

Relocating a labor force skilled in domains different from that prioritized in the region: Such labor is generally more geographically mobile and is likely to relocate to diversified cities or other regions with demand for their skills. Generating local human capital that is fit for the local specialization will be costly and may require time. Local training programs that reinforce the formation of general and specialized skills may further induce emigration of locally produced human capital (McCann and Ortega-Argilés 2011). The question here is whether countries are ready to accommodate these effects.

Smart Specialization: Incomplete Information and the Political Economy of Lobbying

The capacity to obtain and process information related to supply and demand conditions is a classic problem of economic policy making. Information-related problems are larger when the products or services will only exist in the future—as in the case of targeted innovation policy. The task becomes even harder when those products need to generate a "transformational" effect—as expected in the present context. The trickier part here is identifying the sufficient conditions, such as relevant economies of scale and spillover effects based on existing comparative advantages. In the 1990s, the difficulty of identifying sectors with the attributes that would render targeted support justifiable was one of the reasons why strategic trade policies did not develop much as a practical option.

The problem of collecting and processing information on supply and demand, even when market conditions are known, is another classic challenge—this time for economic planning. The best known instrument to collect and process information on cost and preferences and on supply and demand is the market, despite the well-established case of market failures. The problem is even more severe in the case of innovations or entrepreneurial activities that could potentially lead to a transformation in the economy by generating a cluster of specialization. The reason is that these activities, by definition, suffer from the lack of a price and, more generally, from missing markets. Since the product is unknown beforehand, the market cannot be created.

Collective decision making, even when based on a broad consultative process, does not help much, for two reasons. The first is asymmetric information. Self-motivated sector or project advocates naturally possess more information about their own sector or project than do policy makers, and so the quality and relevance of information provided or generated inevitably has limited use for decision making. The second reason is that representation in the collective decision-making process will be biased toward incumbent political and economic interests. Incumbents have more incentive to voice or lobby than do new entrants, particularly future entrepreneurs, simply because these entrepreneurs do not exist at the time of targeted policy making and hence have no lobbying power.

The critical factors when designing a smart specialization strategy are the lack of market-generated information and the lack of any alternative source of reliable information for decision making. This is precisely why Hausmann and Rodrik (2003) argued in favor of subsidizing entry into new markets: the information generated by the first comer is a nonproprietary good that can be quickly used by followers (imitators). This in turn sharply reduces the returns to investments in the discovery process, which makes private investment levels inferior to the socially optimal level (as with R&D investments, for example).[5]

When reliable market information is unavailable, governments should not bet public money on targeting specific sectors. One argument often heard is that private sector investors, including venture capitalists and serial entrepreneurs, also have incomplete information and face high uncertainty, but still profit from those investments. Could the public sector achieve the same results? It is very unlikely—given the different incentive regimes under which the public and private sectors operate.

Still, in some circumstances, the lack of information poses no serious risk to policy makers. We argue that, depending on the amount of market-generated information, it may not be desirable to target research and innovation policies to specific sectors. The argument does not exclude the possibility of targeting sectors when enough market-driven information is available. In these cases, focusing R&D policies on sectors in which the economy specializes and is globally competitive is possible, and may help increase the impact of research and innovation investments (figure 2.1). In the next section, we will develop a tentative typology based on the presence of evidence on a region's economic specialization.

Operationalizing Smart Specialization: Access to Information

The information available on the specialization of the country or region is crucial for the success of targeted policies. In developing RIS3, policy makers will have to address the type of economic specialization, and its determinants or binding constraints. On the basis of these elements, they can decide whether research and innovation policy is able to foster development and, if so, whether targeted instruments are advisable.

The simplified typology presented above relates the degree of information on a region's economic specialization to the convenience of adopting targeted policies (figure 2.2). We propose three cases in which different degrees of information imply different chances of success with sector targeting, and we provide

Figure 2.1 Access to Information, Risk Level, and Policy Making

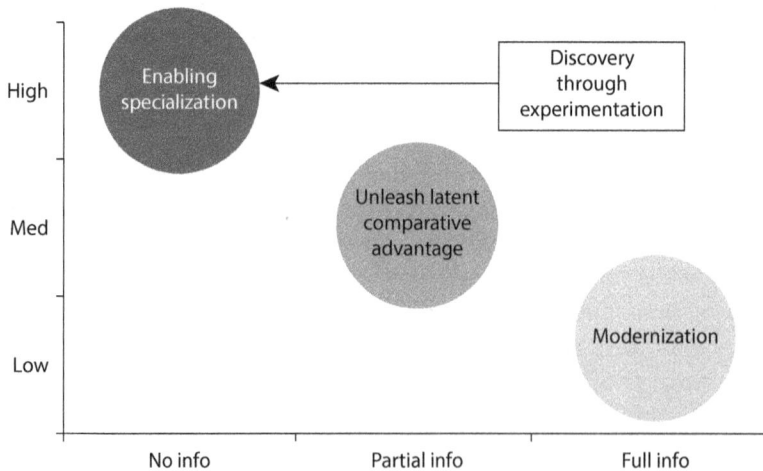

Source: Correa and Guceri 2014.

Figure 2.2 RIS3—A Simplified Typology

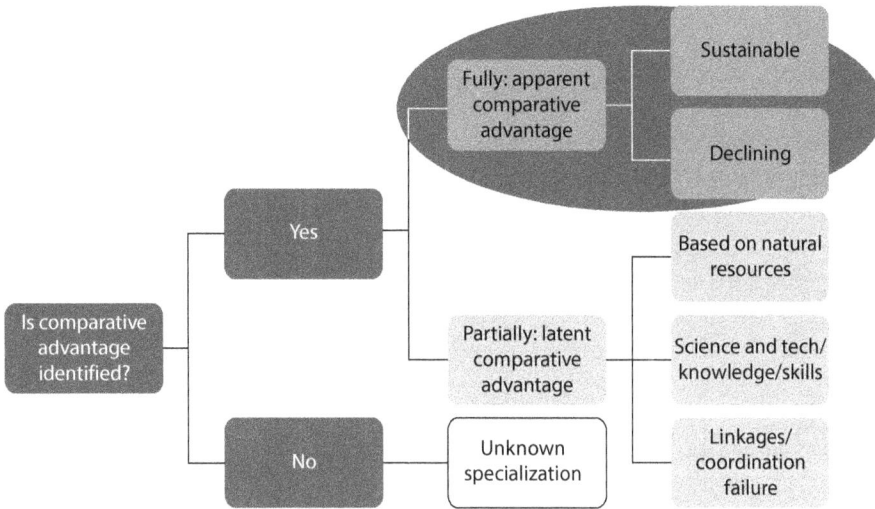

Source: Correa and Guceri 2014.

some practical examples: regions with apparent comparative advantage; regions with latent comparative advantage; and regions with unclear specialization.

- *Regions with apparent comparative advantage.* These are the regions where a few globally competitive industries are already installed. For such regions, the key indicators and consultative processes should lead to similar conclusions on their comparative advantage. Yet regions of this type may be experiencing growth or decline and, consequently, the pattern of specialization may (or may not) be sustainable. Targeted research and innovation policies will complement existing productive assets, helping firms either maintain a competitive edge in the sector by investing in R&D or regain competitive advantage lost to new players in the global market. The RIS3 Guide (European Commission 2012) presents examples: the case of the Finnish pulp and paper industry, which is increasing its R&D efforts in nanotechnology to improve the efficiency and quality of production; and Spanish CITES (public–private laboratories providing technology-related services to the private sector), which were instrumental in improving product design and quality in the local shoe industry, helping the sector refocus exports on high-end markets (particularly Japan).

- *Regions with latent comparative advantage.* Here, the potential area of specialization has no substantial economic activity, but the required knowledge "partially" exists typically (but not only) because of either availability of a nontradable, location-specific input, such as a natural resource or an immovable asset (land and climate, for example), or local common knowledge on economic activity—a tradition prevalent in the region that indicates potential for specialization. Again, assuming that the bottleneck is related to lack of knowledge of local content, research and innovation policies (and probably investments in skills formation and other business-development services) may be useful to "unleash" existing comparative advantage. The cases of the soybean industry in Brazil and of wine in Chile are good illustrations of how technology-oriented policies may enable the development of entire sectors, triggering structural change.

- *Regions with unclear specialization.* When there is no obvious local asset that induces economic specialization in a region, when information from direct observation of market dynamics is unavailable, and when sector targeting becomes less recommendable for the reasons discussed above, we suggest that policy makers focus on creating an enabling environment for efficient market selection (from which an economic specialization should emerge over the years). This means combining measures that promote firm entry and start-up—possibly high-growth potential firms—and that allow firm exit.

The challenge therefore is to define the right policy mix and the programs that better fit the economic and institutional circumstances of each country or region to a predefined objective. Yet predefined entrepreneurship policies do not exist as such; rather, existing economic policies constitute (or not) an ecosystem that is favorable to entrepreneurship (Acs 2012). An ecosystem that promotes entrepreneurship should allow business opportunities to be identified and should provide the right incentives and access to inputs for entrepreneurs to develop them. For instance, inadequate regulation of the telecoms or logistics sectors may preclude the development of e-commerce; in other cases, the lack of early-stage financing (as discussed in chapter 5) may hinder development of science-based start-ups.

Smart Specialization: Implications

Ex ante targeting is more likely to achieve its goals if the apparent or latent comparative advantage of a region is known to the policy maker. In such regions information is available on their specializations, and the risk of moral hazard and opportunistic lobbying is low. In these regions, too, are clear policy implications pointing toward specialization, which can be welfare improving by facilitating the efficient use of available resources.

By contrast, peripheral regions, with less clear emerging or existing trends, need their future specializations to be revealed through a flexible strategy that encompasses enabling policies for entrepreneurship and market selection, rather than ex ante targeting. While agglomeration forces stimulate the creation of a diversified core and a rather specialized periphery, the challenge for the periphery is to create a suitable environment for entrepreneurship. This is difficult because entrepreneurship is more likely to flourish in regions where a diverse set of sectors exists; many multinational companies operate; markets are competitive; the population is dense; and market potential is high (McCann and Ortega-Argilés 2011). In the typically less-developed regions, the prime goals of the policy maker may include facilitating firms' access to information, improving market entry and exit conditions, building the infrastructure for innovation financing, and helping the buildup of knowledge assets (such as a large research university with commercialization potential). Research and innovation policy may focus on commercializing research through university–industry collaboration, launching research start-ups and spin-offs, improving the intellectual property regime, developing the early-stage financing infrastructure, strengthening knowledge networks, and facilitating spillovers.

Still, achieving structural transformation of a region extends beyond research and innovation policies. Trade policy, technology absorption and adoption, the business environment, and the regulations governing human capital input are directly related to the entrepreneurial environment. Policy setting in these areas naturally involves a more complex institutional framework than the boundaries of research and innovation policy. To be more specific, the entrepreneurial environment is shaped by a whole host of policies, including labor market regulation, education policy (strong primary, secondary, tertiary, vocational, and higher education), health policy (which helps provide security for the entrepreneur), the regulatory environment, business taxation, and the judicial system (Acs 2012). Rigid business regulation may deteriorate entry conditions, hindering the region's entrepreneurial capacity.

The government, which has an incentive to support self-discovery, should consider addressing entry barriers. Self-discovery—crucial for targeting activities—depends heavily on entry conditions, with fewer discoveries occurring in systems burdened by barriers to entry. Klinger and Lederman (2011) document the strong correlation between government regulation and self-discovery. The higher the regulatory burden, the fewer export discoveries are made, despite the threat of allowing imitators in the newly discovered sectors. In fact, imitation tends to increase welfare in many new sectors because it increases the scope for social returns from these activities.

The right policy mix is not necessarily related to research and innovation policy, but is one that enables the self-discovery process. The nature of entrepreneurial activity is to experiment with new product niches, discovering the cost of these activities and abandoning those that cannot be undertaken at low enough cost to prove profitable. As Rodrik (2004) elaborates, the discovery of such cost does not necessarily arise from inventing a product or process new to the world (although it may well be). Such discovery may even arise from adopting a technology to produce a traditional product or undertake a well-known process at lower cost. Another route is to adapt the product or process to local conditions more efficiently than the competition.

When designing RIS3, the first issue to be evaluated is whether knowledge is the binding constraint to the structural transformation of the region. If that constraint relates to a more fundamental structural bottleneck, such as the regulatory environment or physical infrastructure, the region should prioritize tackling it, while considering the next steps in research and innovation in its medium- or long-term agenda. For instance, if the business regulatory framework for establishing a company is heavily bound up in red tape, taking measures to generate university spin-offs may prove insufficient to stimulate the desired structural transformation of the local economy.

Having established whether knowledge is a binding constraint, the policy strategy should then sketch more specifically what the binding constraints are, and which ones to address in the short, medium, and long terms. Knowledge bottleneck(s) may relate to a range of institutional factors, usually including constraints to increasing private research and innovation investment (access to finance, intellectual property protection, incentive schemes), constraints to improving technology adoption by small and medium enterprises (SMEs) (skill mismatches, ease of access to technology, awareness), and constraints to research excellence (skilled labor, partnerships).

RIS3 can be designed as a long-term framework to install and continually review the policy approach through integrated impact evaluation and iterative learning. The framework should do the following:

    i. Determine the overall objective and relevant measurable goals.
   ii. Identify bottlenecks and market failures.
  iii. Experiment, learn from these experiments, and adapt policies accordingly.

The "identification stage" consists of tasks i and ii (table 2.1 below), and the "experimentation and adaptation stage" of task iii (figure 2.3 below).

RIS3 identifies channels through which measurable development objectives can be achieved via flexible policy interventions. The monitoring and evaluation framework should therefore be embedded in all stages of RIS3. Progress toward these objectives can be monitored through intermediate goals and influenced via flexible interventions. In other words, the policy maker works backwards from the desired final outcomes, identifying the main channels through which these outcomes can be achieved via interventions.

Rather than ex ante targeting of certain sectors, research and innovation policies could do the following: focus on identifying the intermediate goals, the channels through which these goals are to be achieved, and the bottlenecks—market and institutional failures—that reduce the chances of success. Public actions would then tentatively aim at correcting such failures; be monitored and evaluated; and be adjusted on the basis of lessons systematically learned. A recognition that there are no predefined policy recipes when economic specialization is not immediately evident implies the need to raise the relevance of results orientation and to integrate proper monitoring and evaluation mechanisms with policy and program implementation.

And so, to raise the impact of research and innovation on regional and national development, governments could focus their interventions on a few measurable

Table 2.1 Examples of RIS3 Identification Stage

Intermediate goal	Bottleneck	Instruments	Measure of success
Incentivizing R&D expenditure	Financing	Alternative credit lines	No. of firms engaged in innovation; proportion of spending on R&D
	Incentives	Matching grants	
Accelerating R&D commercialization	Business skills (academia)	Entrepreneurial training	Licensing of spin-off companies; patents
	Early-stage financing	Angel investors	
Improving technology adoption by SMEs	Awareness	Information campaigns	Take-up rates of targeted technologies
	Skills	Training programs	
Achieving research excellence	Retention of skilled labor force	Nonfinancial incentive schemes	Academic qualifications; quality of publication output; publication collaborations
	Coordination and collaboration	Workshops and international conferences	

Source: Correa and Guceri 2014.

intermediate goals (table 2.1) with a clear link to the broader development aim of higher growth and job creation, by the following:

- Incentivizing business expenditure in research and innovation, including new business models and the number of firms engaged in innovation activities.
- Accelerating the commercialization of public research and academic entrepreneurship through licensing and spin-off companies.
- Enabling faster adoption of updated technologies and organization processes, especially general-purpose technologies by SMEs and ICT by the services sector.
- Improving research excellence by increasing collaboration between public research organizations with the local private sector through organic research collaboration, and by reforming the system of managing public research organizations to favor research excellence and productivity, as well as skills formation.

Selected research and innovation interventions would therefore directly contribute to one of the previously selected intermediate goals and be evaluated accordingly. They would address identified market and institutional failures and be designed and implemented in a way consistent with robust evaluation, thus informing subsequent adjustment toward greater impact. Such an approach would particularly help rationalize the relevance of investments in research infrastructure (important for the use of European Structural and Investment Funds).

- With commercialization of public research, for instance, regions could experiment with programs addressing the supply of ideas originating in public research organizations, from supporting better intellectual property

Figure 2.3 Examples of RIS3 Experimentation and Adaptation Stage

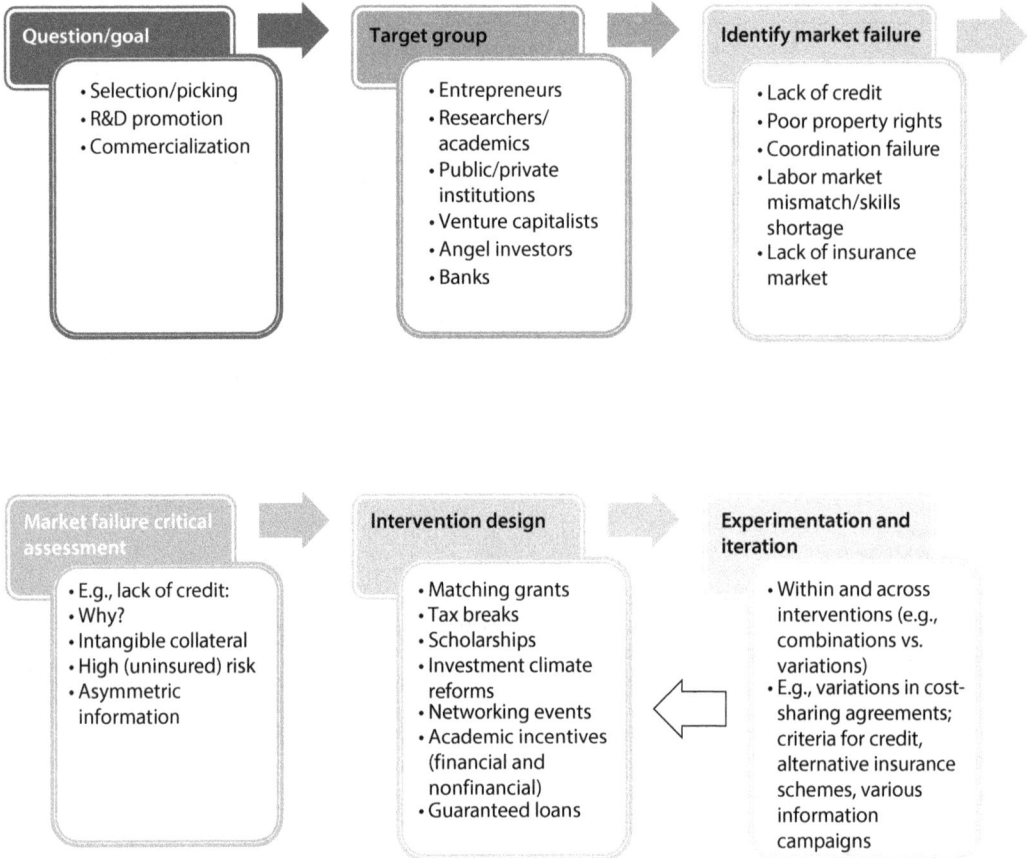

Question/goal
- Selection/picking
- R&D promotion
- Commercialization

Target group
- Entrepreneurs
- Researchers/academics
- Public/private institutions
- Venture capitalists
- Angel investors
- Banks

Identify market failure
- Lack of credit
- Poor property rights
- Coordination failure
- Labor market mismatch/skills shortage
- Lack of insurance market

Market failure critical assessment
- E.g., lack of credit:
- Why?
- Intangible collateral
- High (uninsured) risk
- Asymmetric information

Intervention design
- Matching grants
- Tax breaks
- Scholarships
- Investment climate reforms
- Networking events
- Academic incentives (financial and nonfinancial)
- Guaranteed loans

Experimentation and iteration
- Within and across interventions (e.g., combinations vs. variations)
- E.g., variations in cost-sharing agreements; criteria for credit, alternative insurance schemes, various information campaigns

Source: Correa and Guceri 2014.

management practices by technology transfer offices to providing small grants for developing proof of concepts; they could also experiment with, for example, "accelerator" programs.

- With ICT adoption, the lower use by European firms than by their U.S. peers derives from relatively lower returns from ICT investments in Europe due to the adoption of less efficient management (organizational) practices and stiffer product market regulation, which prevent local firms from reaping the same benefits as their U.S. counterparts (Bloom, Sadun, and van Reenen 2012).[6] Hence there is no obvious indication of whether policies (and which type) could help increase ICT productivity and increase ICT use in Europe.

- Experimentation may also be related to the design of the support mechanism (figure 2.3). For instance, in promoting collaboration between public research organizations and the private sector, one of the important variables may be the targeted population. Under certain conditions—difficult to identify up front—that eliminate the possibility that the private sector will sponsor project applications may leave a large share of funds unused (especially when other standard sources of public financing for public research are available).

In Croatia itself, the assessment of innovation policies is still in its initial stages. The above framework could help the government better deploy additional resources to promote research and innovation. It may also help it establish a strategy to achieve related goals set under its innovation strategies. Finally, it should help the government identify remaining areas of critical institutional reforms to complete the transition of the innovation systems.

Promising Sectors for Smart Specialization

Weaving the above implications into the reality of Croatia today, four sector case studies illustrate possible RIS3 approaches. They do not imply these activities are priority sectors but hold them up as promising areas of potential research and innovation specialization at local level. Outlined in appendix A, they cover clean energy, oyster production, Slavonski kulen (a traditional meat product), and biotechnology and pharmaceuticals.

Notes

1. http://s3platform.jrc.ec.europa.eu/s3pguide.
2. Inputs for this section are based on Correa and Guceri (2014).
3. COM (2010) 2020.
4. European Commission Europe 2020 Flagship Initiative Innovation Union.
5. Hausmann and Rodrik (2003, 608) justified policy intervention in the case where entrepreneurs are searching for their ideal domain of operation under uncertainty about costs, which depend on some unobserved productivity parameter for each good in the "modern" sector. They solve for the general equilibrium investment in the modern sector and show that the decentralized equilibrium results in underinvestment, coupled with too high product diversification.
6. European Commission (2010b) argued that other reasons for the gap are low investment in ICT equipment, which remains below the threshold that could potentially allow the private sector to reap the benefits of ICT investment, as well as too little progress in innovative activity by European businesses and inadequate research infrastructure to realize the productivity gains from ICT investment. If the underlying reason is below-threshold ICT investment, the reinforcement suggested by the EU could help.

Bibliography

Acs, Z. 2012. "Public Policies in an Entrepreneurial Society." In *Handbook on the Economics and Theory of the Firm*, edited by M. Dietrich and J. Krafft, 515–28. Cheltenham, UK: Edward Elgar.

Bloom, N., R. Sadun, and J. van Reenen. 2012. "Americans Do IT Better: US Multinationals and the Productivity Miracle." *American Economic Review* 102 (1): 167–201.

Correa, P., and I. Guceri. 2014. "Research and Innovation for Smart Specialization Strategy: Concept, Implementation Challenges and Implications." World Bank, Washington, DC.

David, P., D. Foray, and B. Hall. 2009. "Smart Specialisation—The Concept." Knowledge Economists Policy Brief 9, European Commission, Brussels.

Duranton, G., and D. Puga. 2004. "Micro-Foundations of Urban Agglomeration Economies." In *Handbook of Regional and Urban Economics*, edited by J. V. Henderson and J. Thisse, 2063–117. Amsterdam: Elsevier.

European Commission. 2010a. "Europe 2020: A Strategy for Smart, Sustainable and Inclusive Growth." Communication from the Commission, COM (2010) 2020, Brussels.

———. 2010b. "Europe 2020 Flagship Initiative Innovation Union." Communication from the Commission to the European Parliament, the Council, the European Economic and Social Committee and the Committee of the Regions, COM (2010) 546, Brussels.

———. 2012. *Guide to Research and Innovation Strategies for Smart Specialisation (RIS 3)*. Luxembourg: European Union. http://s3platform.jrc.ed.europa.eu/s3pguide.

Foray, D., and B. van Ark. 2007. "Smart Specialization in a Truly Integrated Research Area Is the Key to Attracting More R&D to Europe." Knowledge Economists Policy Brief 1, European Commission, Brussels.

Hausmann, R., and D. Rodrik. 2003. "Economic Development as Self-Discovery." *Journal of Development Economics* 72 (2): 603–33.

Klinger, B., and D. Lederman. 2011. "Export Discoveries, Diversification and Barriers to Entry." *Economic Systems* 35 (1): 64–83.

Krugman, P. R. 1991. "Increasing Returns and Economic Geography." *Journal of Political Economy* 99 (3): 483–99.

McCann, P., and R. Ortega-Argilés. 2011. "Smart Specialisation, Regional Growth and Applications to EU Cohesion Policy." Economic Geography Working Paper, Faculty of Spatial Sciences, University of Groningen, the Netherlands.

Rodrik, D. 2004. "Industrial Policy for the Twenty-First Century." Centre for Economic Policy Research Discussion Paper Series 4767, Harvard University, Cambridge, MA.

CHAPTER 3

Croatia's Trade: Performance, Competitiveness, and Potential

Chapter Summary[1]

Croatia's trade competitiveness gives reasons for hope—and for concern. On the positive side, first, the country managed to penetrate emerging and fast-growth regional markets such as the EU-12, the Russian Federation, and the Middle East and North Africa. Second, some technologically advanced niche product categories in traditional sectors and in new activities are emerging. Third, the country has developed a small set of strong, multiproduct exporters as well as regional multinationals to serve the former Yugoslav market. Fourth, it has developed farm-to-retail supply chains in food processing. Finally, Croatia has a supportive soft and hard infrastructure, notably its highly trained technical workforce in, for example, biotechnology and engineering, ICT, and transport.

Still, its overall export competitiveness (as measured by export market share) is declining. The decline is driven by negative pull and push factors. But while Croatia and all its peers suffered from unfavorable sector and geographic specialization (pull factors), only Croatia showed a negative supply-side contribution (push factors) to growth in export market share. Unfavorable domestic elements became the determining factor in the loss of export market share after the onset of the global crisis.

Croatia faces notable challenges in increasing its competitiveness in sophisticated regional markets and in sustaining the growth of complex products (pull factors). It lost export relevance in the EU-15 market because of increased market saturation and severe export competition. Croatia must overcome its mediocre performance on export sophistication: despite promising instances in niche products and sectors, the overall complexity and sophistication of its goods and services export basket have only shown modest gains over the past decade. In fact, the country is trailing all its regional peers on these two metrics. It also needs to sustain the growth of its emerging sophisticated niche products, but this alone is insufficient: greater sophistication and value addition require a larger share of complex products to be among the top exports.

Croatia's emerging niche product categories should not be made the focus of any strategy to "pick winners," but indicate the need for horizontal policies that support entrepreneurship and market discovery. Most of the new, growing sectors are still isolated in the product space (i.e., they neither form new clusters of economic activity nor represent expansion of existing ones)—and in principle even represent technological discontinuity.

Having been in recession for the past five years, the export performance of Croatia—a small country of some 4.3 million—is critical. The channels through which export expansion enhances productivity and growth are well known: exports allow for specialization in a country's comparative advantage and thereby raise growth. Export orientation also includes dynamic efficiency gains in the more productive export sector engineered by higher competition, greater economies of scale, better capacity utilization, dissemination of knowledge and technological progress, and improved allocation of scarce resources throughout the economy. Moreover, higher exports are linked to productivity gains that should lead to wage increases when factors are paid their marginal product. Finally, increased export earnings relax current account pressures by improving the odds of importing the necessary intermediate and capital goods and of attracting foreign investment, stimulating growth through capital accumulation (Medina-Smith 2001; Mahadevan 2007). What, then, has been Croatia's recent pattern of export performance?

Export Growth

Croatia has shown uninspiring trade and development performance since the onset of the global crisis (figure 3.1). Its export openness and gross domestic product (GDP) per capita stagnated between 2007–08 and 2011–12, while some comparator countries, especially Estonia, Hungary, Latvia, and Lithuania, sharply raised their share of exports in GDP. The country's export openness still lies below the average of countries with similar incomes, with its export share below the regression line, suggesting that openness to trade is lower than expected given the country's per capita income. Only Poland finds itself in a similar position.

Croatia's export growth has been virtually stagnant, while export competitiveness as measured by export market share changes has even declined. Exports grew at 2.5 percent on average a year over 2006Q1–2013Q1, but this performance was driven by the precrisis period. Croatia exhibited the slowest export growth, against double-digit growth in Poland and the Slovak Republic, and strong growth in Hungary and Slovenia (table 3.1). Croatia's export market share declined by an annual 4.7 percent over the period (figure 3.2). (Changes in world export market shares are often used by policy analysts as a main indicator and common way to measure export competitiveness.)

Croatia's declining export competitiveness is driven by pull and push factors (see table 3.1). Pull factors reflect market and sector export composition; push factors describe a country's own supply-side capacity to expand export market shares, assuming equal market and sector export composition across all countries.

Figure 3.1 Export Openness and per Capita GDP, 2007–08 vs. 2011–12

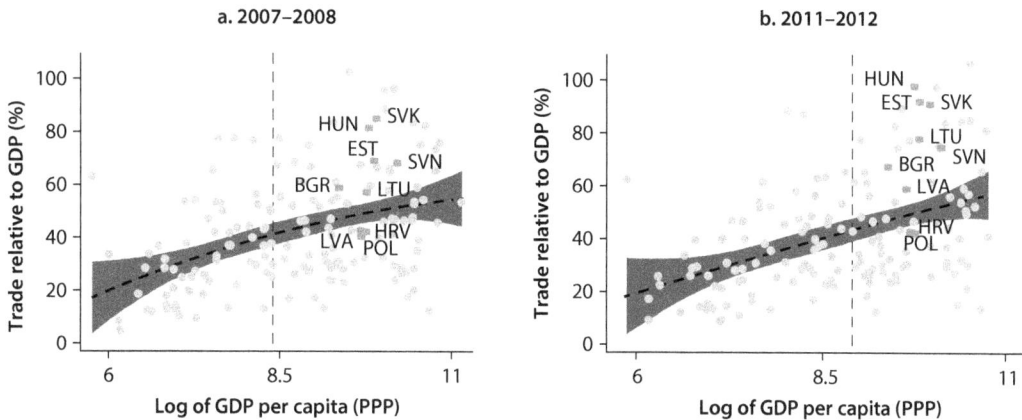

Sources: World Development Indicators; UN Comtrade.
Note: The figure compares exports as a share of GDP with per capita GDP for Croatia and its comparator countries. It also includes the 95 percent confidence intervals around the fitted line to show to what extent a country is an outlier relative to the prediction. The upward-sloping line indicates the fitted values obtained by a linear prediction of the relationship between the two variables, and reflects the stylized fact that exports represent a larger share of GDP for richer countries.

Table 3.1 Export Market Share Growth Decomposition across Five Eastern European Countries, 2006Q1–2013Q1

			Pull factors		Push factors		
	Export growth	Export market share change	Geographic	Sector	Values (overall)	Price	Volumes
Croatia	**2.5**	**−4.7**	**−2.9**	**−0.9**	**−0.9**	**−0.5**	**−0.4**
Hungary	8.1	0.9	−0.9	−1.0	2.8	0.0	2.8
Poland	13.4	6.2	−1.1	−0.6	7.9	−0.9	8.7
Slovak Republic	13.3	6.1	−1.0	−1.3	8.5	0.3	8.2
Slovenia	7.5	0.3	−1.8	−0.8	2.8	−0.9	3.7

Source: Export Competitiveness Database.
Note: Indicators are expressed in log-difference form, which allows for additivity across indicators.

Market Orientation and Export Diversification

The regional distribution of exports has changed little over the past decade, but Croatia's main export partners in Europe show some shift. Over 2002–12, other European countries remained Croatia's main export destination (figure 3.3), showing only a slight decline over the period from 86 percent to 80 percent of total exports. However, the market composition within Europe changed: while the traditional block of the 15 older EU members (EU-15) is still important, their export share fell from 53 percent to 41 percent, explained by increased market saturation and severe export competition. The export share to the new EU-12 member countries, by contrast, increased from 13 percent

Figure 3.2 Export Growth and Change in Market Share, 2006Q1–2013Q1

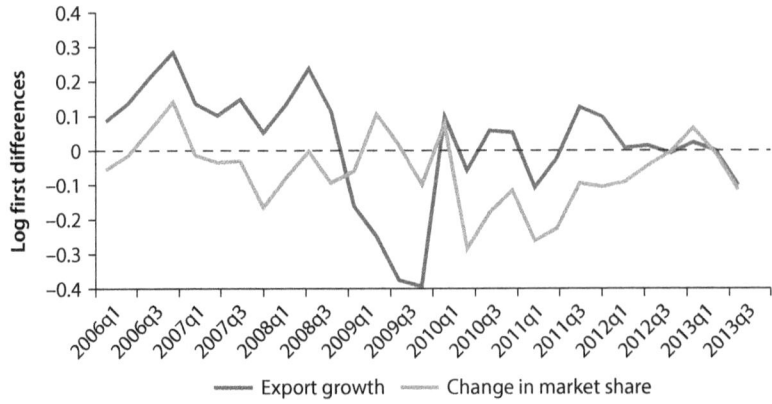

Source: Export Competitiveness Database based on Gaulier et al. (2013).
Note: Indicators are expressed in log-difference form.

Figure 3.3 Export Market Distribution, Croatia, 2002 vs. 2012
Percent

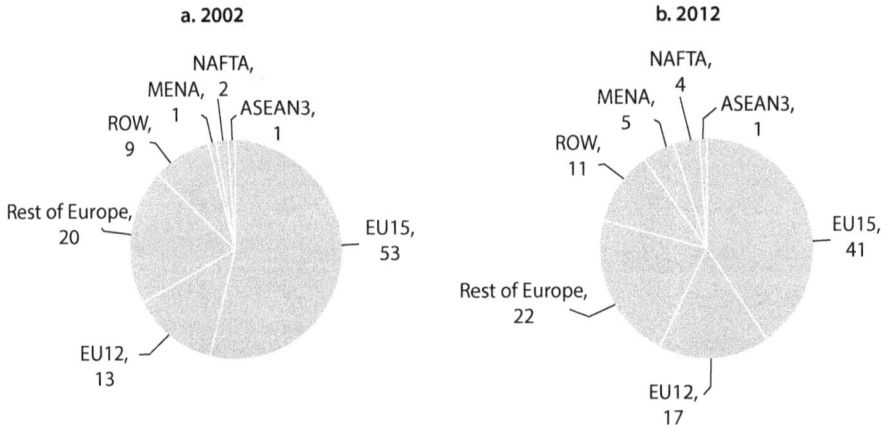

Source: Gaulier et al. 2013.

to 17 percent. According to field interviews, their growing share stems from easier market access, and by similar tastes and business practices. The EU-12 members are also characterized by higher growth, an expanding middle class, and growing purchasing power, but still demand less sophisticated products than the EU-15. Croatia's export share to non-EU countries in Europe edged up from 20 percent to 22 percent for reasons similar to those of the EU-12, as well as previous economic links. Croatia's export share to the Middle East and North Africa and to countries in the North American Free Trade Agreement (NAFTA) tripled from 3 percent to 9 percent. Croatia's export markets were more diversified in 2012 than a decade earlier, mainly due to the declining export share of the EU-15 markets.

All Croatia's regional peers except Lithuania diversified their export markets. Among them, Croatia shows the second-lowest export market concentration after Bulgaria, down from third highest a decade earlier, as seen in figure 3.4, which shows market diversification per the Herfindahl-Hirschman Index, a commonly used measure of market concentration. Figure 3.5 shows

Figure 3.4　Market Concentration: Herfindahl-Hirschman Index, Croatia and Peer Countries, 2002 vs. 2012

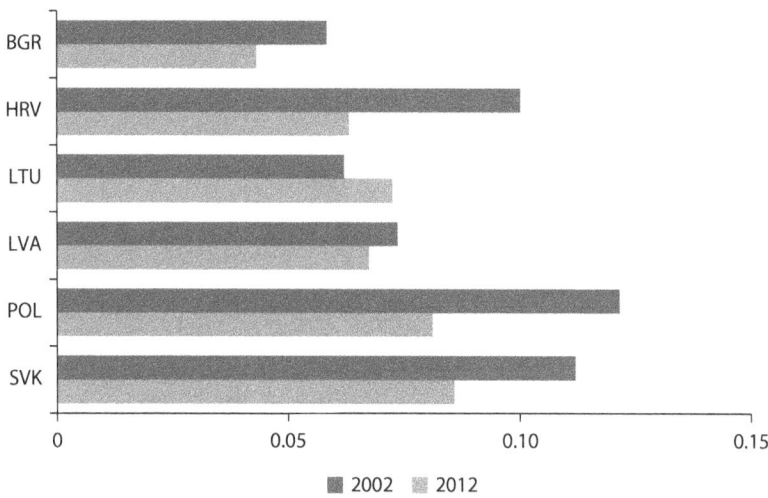

Figure 3.5　Export Growth Orientation, Croatia's Top 10 Export Markets, 2007–12

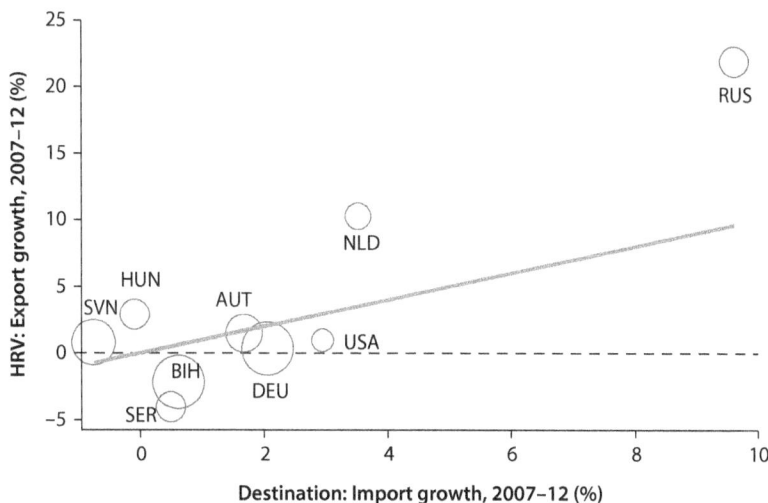

Croatia's top 10 export markets: its export growth exceeded import growth in Slovenia, Russia, Hungary, and the Netherlands, where it gained market share, while it lost market share in Bosnia and Herzegovina, Germany, Serbia, and the United States.

Five sectors covered about two-thirds of Croatia's export basket in 2012, when it sold products to 77 export markets. The five were minerals and mineral fuels (15.1 percent), metals (10.8 percent), chemicals (9.7 percent), machinery (17.9 percent), and transport goods (9.1 percent). All these sectors, except transport goods, posted annual double-digit growth in the past decade (table 3.2). In other sectors, only foodstuffs and wood reported export values close to US$1 billion. Textiles declined from 11.7 percent to 4.9 percent of total exports.

Growth in the broad sectors in which Croatia specializes is subdued. Figure 3.6 compares its top 20 export sectors with worldwide exports of the same sectors. Products in the bottom-right quadrant represent products in which Croatia expanded its market share in world exports, but whose world import demand growth increased at a relatively low rate over 2008–12 ("winners in declining sectors"). These include, in particular, iron and steel, fertilizers, and pharmaceutical products—the sector with the largest export value. Croatia had several important sectors (by overall export value) that both lost market share in world exports and faced lower import demand ("losers in declining sectors," bottom-left quadrant), including clothing/textiles, wood products, articles of iron, steel, aluminum products, vehicles, ships, and boats and other floating structures. Many of these sectors have

Table 3.2 Croatia's Exports, 2002–12

	US$ million		Total exports (%)		RCA		Annual growth rate
	2002	2012	2002	2012	2002	2012	
01–05 Animal	102.7	287.7	2.1	2.3	1.0	1.4	10.9
06–15 Vegetable	87.2	361.0	1.8	2.9	0.6	0.9	15.3
16–24 Foodstuffs	370.4	944.5	7.6	7.6	2.4	2.6	9.8
25–27 Minerals	559.0	1,862.7	11.4	15.1	1.1	0.7	12.8
28–38 Chemicals	374.0	1,204.4	7.6	9.7	0.8	1.1	12.4
39–40 Plastic/rubber	188.0	268.6	3.8	2.2	0.9	0.5	3.6
41–43 Hides, skins	78.1	174.3	1.6	1.4	1.9	2.4	8.4
44–49 Wood	356.7	852.4	7.3	6.9	2.0	3.1	9.1
50–63 Textiles, clothing	575.4	605.5	11.7	4.9	2.0	1.3	0.5
64–67 Footwear	152.2	207.1	3.1	1.7	3.1	2.3	3.1
68–70 Stone/glass	124.7	297.5	2.5	2.4	2.3	2.8	9.1
71–83 Metals	295.3	1,339.6	6.0	10.8	0.7	1.1	16.3
84–85 Mach/elec	714.5	2,210.6	14.6	17.9	0.5	0.7	12.0
86–89 Transportation	689.1	1,128.5	14.1	9.1	1.2	1.1	5.1
90–97 Miscellaneous	236.3	616.6	4.8	5.0	0.8	0.9	10.1
Total	4,903.6	12,361.0	100.0	100.0	–	–	9.7

Figure 3.6 Growth Orientation of Top 20 Export Sectors, Croatia, 2008–12

Growth of national supply and international demand
for export products of Croatia-2012

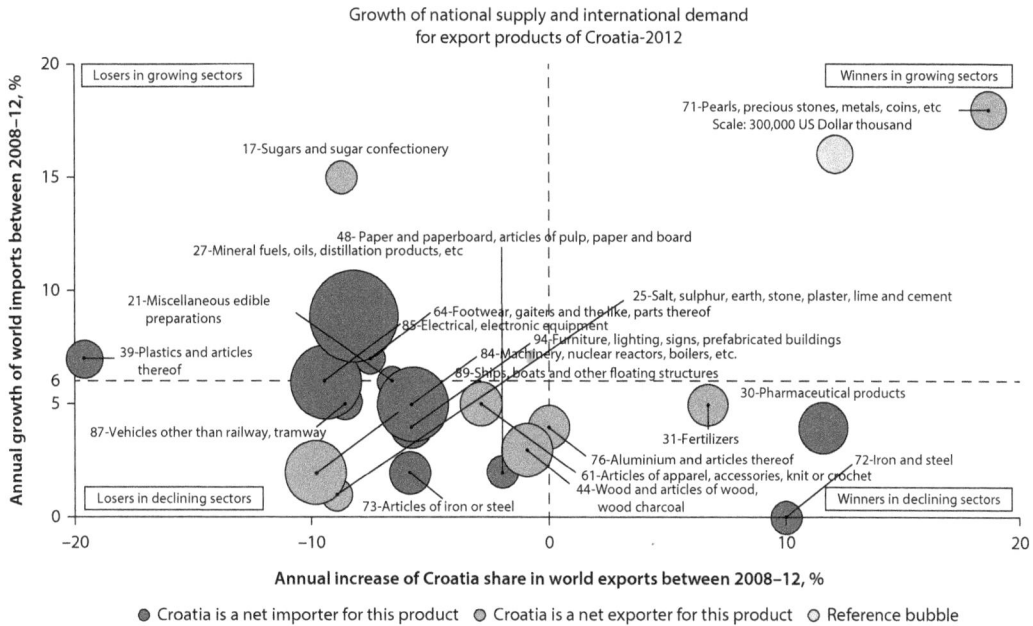

● Croatia is a net importer for this product   ◐ Croatia is a net exporter for this product   ○ Reference bubble

Source: International Trade Centre-Market Analysis Tools.
Note: The y-axis plots annual growth of world imports for Croatia's top 20 export sectors between 2008 and 2012, while the x-axis shows the annual increase of Croatia's export share in world exports for these sectors over the period. The circle size indicates the sector's export value.

large export values, which raises some concern. Equally unfortunately, Croatia expanded only very little its share in world exports for products whose import demand over the period increased strongly ("winners in growing sectors," top-right quadrant). Croatia's exports in this category included only one sector at the HS-2-digit level: pearls, precious stones, and so on. Finally, Croatia had a few products—footwear, sugars and sugar confectionery, mineral fuels, plastic products, and other food—with growing world demand and declining world market export shares ("losers in growing sectors," top-left quadrant).

Crucially, at the Harmonized System (HS) 6-digit level, of the 3,407 products that Croatia exported in 2012 only four were winners in growing sectors. These were medicaments of other antibiotics, for retail sale; cane of beet sugar in solid form, N.E.S.; articles of leather or of composition leather; and revolvers and pistols.

The above findings suggest that Croatia is in a transition phase, specializing in existing markets but showing little growth in new products and markets. The contribution of old products in old markets to export growth is higher than the contribution of old products in new markets, new products in old markets, and, vitally, new products in new markets (figure 3.7).

Figure 3.7 Intensive and Extensive Margin of Trade, Croatia

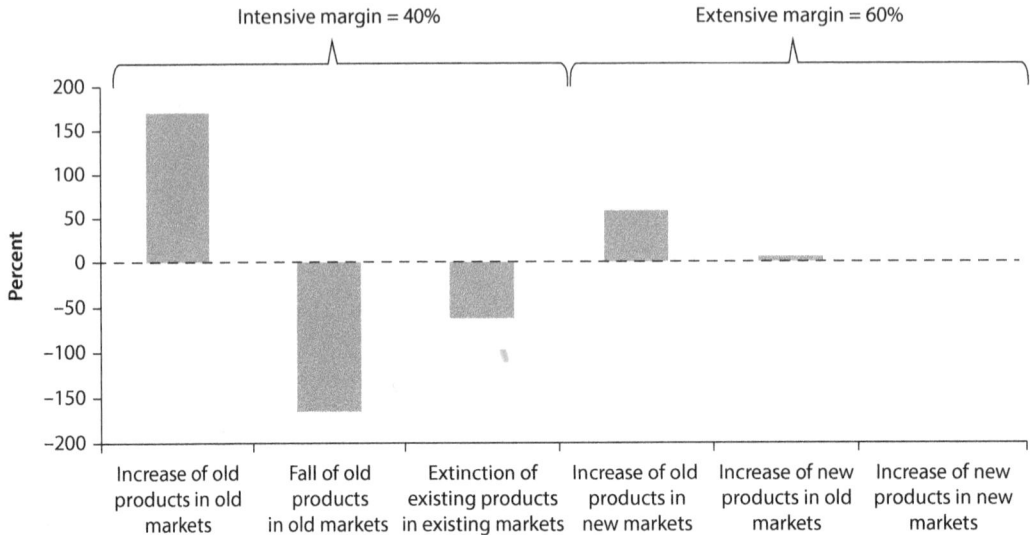

Exports' Economic Complexity, Sophistication, and Quality Upgrading

Croatia's export-basket economic complexity has grown over the past decade but still lags behind peers in the region (figure 3.8). Although Croatia's Economic Complexity Index increased from 0.7 in 2000 to 1 in 2008, it has yet to match those of peers like Hungary, Poland, the Slovak Republic, and Slovenia. Economic complexity is expressed in the composition of a country's productive output and reflects the structures that emerge to hold and combine knowledge. The more varied and useful is the knowledge—or capabilities—embedded in an economy, the more complex it is. The premise of this measure is that some goods (such as jet engines) require large amounts of knowledge possessed by large networks of people and organizations, while others (such as sugarcane) require a narrow knowledge base. More complex economies, reflecting a larger mix of complex products, are better able to deploy the sets of organizational capabilities that a manufacturer needs—from design, marketing, and finance to technology, operations, and law (Hidalgo et al. 2011).

Many of Croatia's exports are low to medium in complexity and sophistication. The main reason for this, and why Croatia lags behind regional peers on complexity, is compositional: a large proportion of the main exports are goods that, by global standards, are not very complex. Table 3.3 lists the country's top 20 exports at the Standard International Trade Classification (SITC) 4-digit level for 2011–12.[2] None of Croatia's most important 15 export sectors ranks in the top decile by economic complexity. Of the nearly 200 products with revealed comparative advantage (RCA) greater than one, about half are clustered around the middle deciles (ranking 244–540 out of 786).

Figure 3.8 Economic Complexity of the Export Basket, Croatia vs. Peers, 2000–08

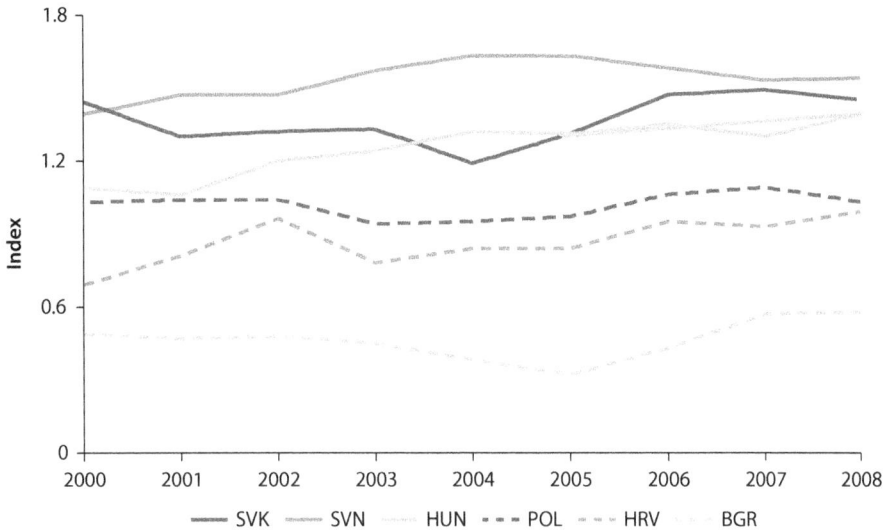

Source: The Observatory of Economic Complexity.
Note: Index available only until 2008.

Table 3.3 Croatia's Top Exports and Their Product Complexity Rank

	SITC	Export	Share in 2011–12 (%)	Complexity rank
1	7932	Ships and boats	7.21	272
2	5417	Medicaments	3.79	161
3	7711	Electrical transformers	2.32	363
4	2483	Worked wood of non-coniferous	2.18	687
5	6842	Worked aluminum and aluminum alloys	1.89	317
6	5621	Nitrogenous fertilizers	1.73	641
7	7938	Special floating structures	1.68	631
8	8211	Chairs and seats	1.67	318
9	2820	Iron and steel waste	1.62	542
10	612	Refined sugar	1.40	599
11	7731	Electric wire	1.35	441
12	980	Edible products N.E.S.	1.31	424
13	8510	Footwear	1.26	541
14	9710	Gold, nonmonetary	1.16	756
15	5629	Fertilizers	1.15	597
16	7284	Specialized industry machinery and parts N.E.S.	1.10	1
17	7849	Other vehicles parts	1.09	62
18	6612	Cement	1.02	648
19	2882	Other non-ferrous base metals	1.00	529
20	7721	Switchboards, relays, and fuses	0.96	127

Source: UN Comtrade and the Observatory of Economic Complexity.
Note: Products ranked on complexity from 1 to 786 (1 being the most complex).

Some emerging exports, however, show promise. The 16th largest export sector in recent years—specialized industrial machinery and parts (SITC 7284)—ranks among the most complex 100 sectors. Accounting for 1.1 percent of total exports in 2011–12, it is also Croatia's most complex export sector. Other sizable exports that are moderately high in complexity include vehicle parts (SITC 7849), contributing 1.1 percent and ranked 62 in complexity; medicaments (SITC 5417), contributing 3.8 percent and ranked 161; and switchboards, relays, and fuses (SITC 7721), contributing 0.96 percent and ranked 127. These exports are "emerging" because they grew in size and significance over the past 10 years, starting from a low base. All told, these results indicate that Croatia is evolving toward a more complex export basket—but to reach overall levels of far greater sophistication and value addition, it will have to raise sharply its proportion of complex products.

Croatia's largest export sectors are low in physical and human capital. This is quantified by the revealed factor intensity of products classified at a more disaggregated level (HS 6-digit),[3] as computed by Shirotori, Tumurchudur, and Cadot (2010).[4] In 2012, only a few of Croatia's major exports used capital that was above what the median export embodied on either dimension, but many of the biggest exports had low physical and human capital content—figure 3.9,

Figure 3.9 Revealed Factor Intensity of Croatia's Exports, 2012

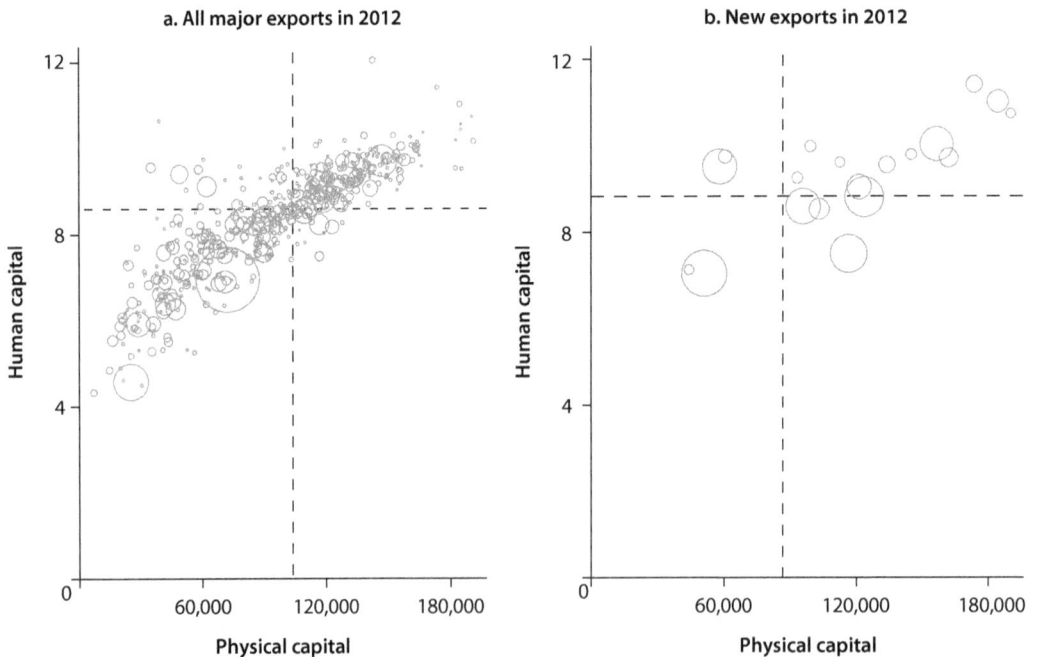

Source: RFI data from UNCTAD.
Note: The left panel plots products based on the relationship between human capital (y-axis) and physical capital (x-axis). The size of the bubble represents the share (importance) of the product in total Croatian exports in 2011–12. The dashed reference lines indicate median values of revealed human and physical capital for individual exports (with a share of at least 0.01 percent in total).

left panel.[5] The panel on the right, however, shows that there is some dynamism among new exports; that is, exports active in 2011–12 but not 2001–02 are sizable, with most also registering a high degree of capital intensity. Further, if a nation's endowment point is represented by the intersection of its average stock of physical and human capital, we can assess how far or close to this endowment point is the human and physical capital content of each export. If those capital coordinates are close enough to the endowment point, it can be inferred that the capital required in producing that good is in line with the country's average achievements in human and physical capital accumulation. From this perspective, the emerging major exports (with high capital content) do not appear misaligned with Croatia's comparative advantage.

Croatian manufacturing is yet to be dominated by industries where quality competition is more important than price competition. Using the taxonomy of Aiginger (2001), we classify exports belonging to industries that have high, medium, and low relative quality elasticity (RQE). Industries with high RQE compete on quality, whereas those with low RQE compete on price.[6] The intuition is that in quality-elastic industries, high price does not necessarily lead to low quantities sold because those products embody intangible traits like design, service, and reliability. Price-elastic industries are those where low costs lead to more exports and higher prices lead to fewer exports. Countries with a large share in high RQE industries have moved away from industries in which low prices are important for competitiveness, and have achieved inter-industry quality upgrading.

Croatia's share of quality-dominated (high RQE) export industries is less than the share dominated by price competition (low RQE) (figure 3.10). Further, the quality-dominated ratio is much higher for all other peer

Figure 3.10 Relative Quality Elasticity, Croatia vs. Peers, 2012

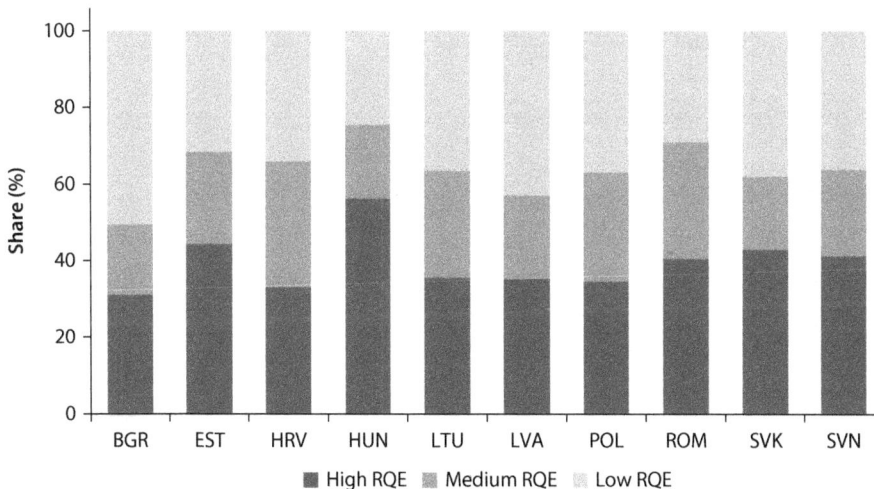

Source: Authors' regrouping of UN Comtrade data based on Aiginger (2001).

countries except Bulgaria (figure 3.11). This is yet another indicator that Croatia has much catching up to do. However, it does have the highest share of medium-RQE industries among its peers, which may help it transition to quality-dominated manufactures. Just like the breakthrough in some pharmaceutical and industrial machinery exports, with medicaments (SITC 5417) and electrical transformers (SITC 7711) doubling their share in a decade, many emerging exports have the potential to be differentiated and made more quality responsive.

One way a country can increase the absolute amount of exports per capita is by augmenting the quality of exports and thus the value of exports per unit. Using a highly detailed database on unit values of exports to the EU (at HS 6-digit level), we construct a measure of the relative quality of each product exported to the EU from countries worldwide. (Relative quality of an individual export is measured by its unit value normalized by the 90th percentile obtained from a distribution of unit values of the same product imported into the EU from all other countries.)

Croatia has made better inroads in sophisticated manufactures than in primary exports to the EU-15, while augmenting their quality. Figure 3.12 shows the change in the relative quality of Croatia's 37 top exports to the EU-15 (accounting for at least 0.5 percent of total exports to that market at HS 6-digit level)[7] and the change in their market share. Nine of the 14 products in the category of "machinery, electronics, and transport equipment" increased both market share and relative quality. This includes photosensitive semiconductor devices (HS 854140), liquid dielectric transformers (HS 850423), and motorboats (HS 890392). One major manufactured export that improved in relative

Figure 3.11 Price vs. Quality Elasticity, Croatia vs. Peers, 2012

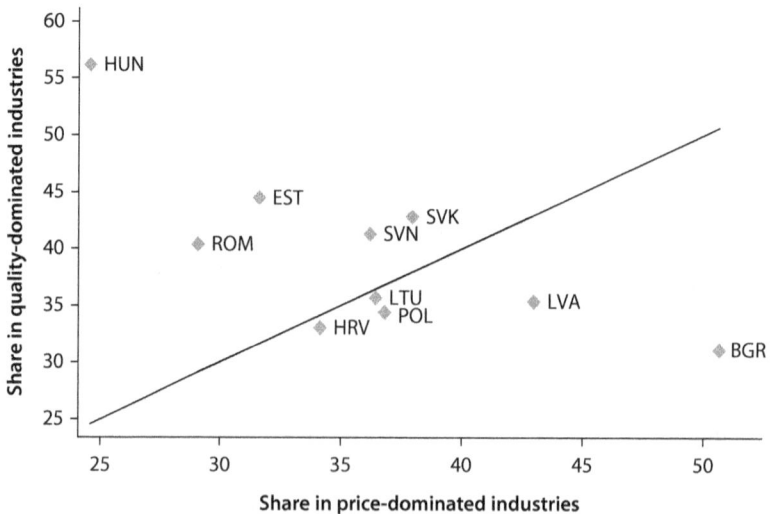

Source: Authors' regrouping of UN Comtrade data based on Aiginger (2001).

Figure 3.12 Change in Relative Quality and Market Share of Top Croatian Exports, 2000–10

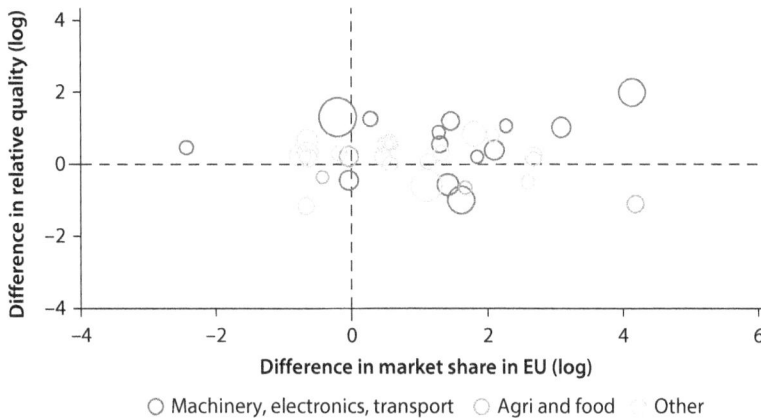

Source: Author's calculation based on CEPII & UN Comtrade.
Note: Europe consists of 15 countries before EU enlargement of 2004. Each dot is a product at the HS 6-digit level weighted by its share in EU in 2010.

quality but lost market share in the EU is "other vessels for transport" (HS 890190). None of the top four exports from the primary sector (agriculture, food and beverages, and wood) improved *both* market share and relative quality; only one—sawn wood of red oak (HS 440791)—improved its relative quality but lost market share in the EU-15.

Product-Space Analysis

The product-space map of Croatia reveals a moderate level of structural transformation over the past two decades. Using tools pioneered by Hidalgo et al. (2007), figure 3.13 presents the product-space map for 1992–93, 2001–02, and 2011–12. The black square dots represent exports in which Croatia had RCA in those years, showing close to 800 tradable products, revealing that over the past decades the cluster of exports around apparel/textiles has lost comparative advantage, while new products are emerging in manufacturing, especially centered on industrial and electrical machinery. Giving more detail, table 3.4 shows that the number of apparel and footwear exports with RCA decreased from 21 to 13 over the last 10 years; similarly, the number of exports with RCA related to the chemicals industry dropped by half, from 24 to 12. In contrast, the number of products with RCA increased the most in the higher-technology machinery sector, from 15 to 43. Although the top export has remained "ships, boats, other vessels" (SITC 7932), the churning is reflected in the shifting significance of top exports. Sawn wood, certain apparel and footwear, and aluminum alloys have become less important now than they were 10 or 20 years ago, overtaken by products like medicaments (SITC 5417) and electrical transformers (SITC 7711).

Figure 3.13 Croatia's Product-Space Maps, 1992–2012

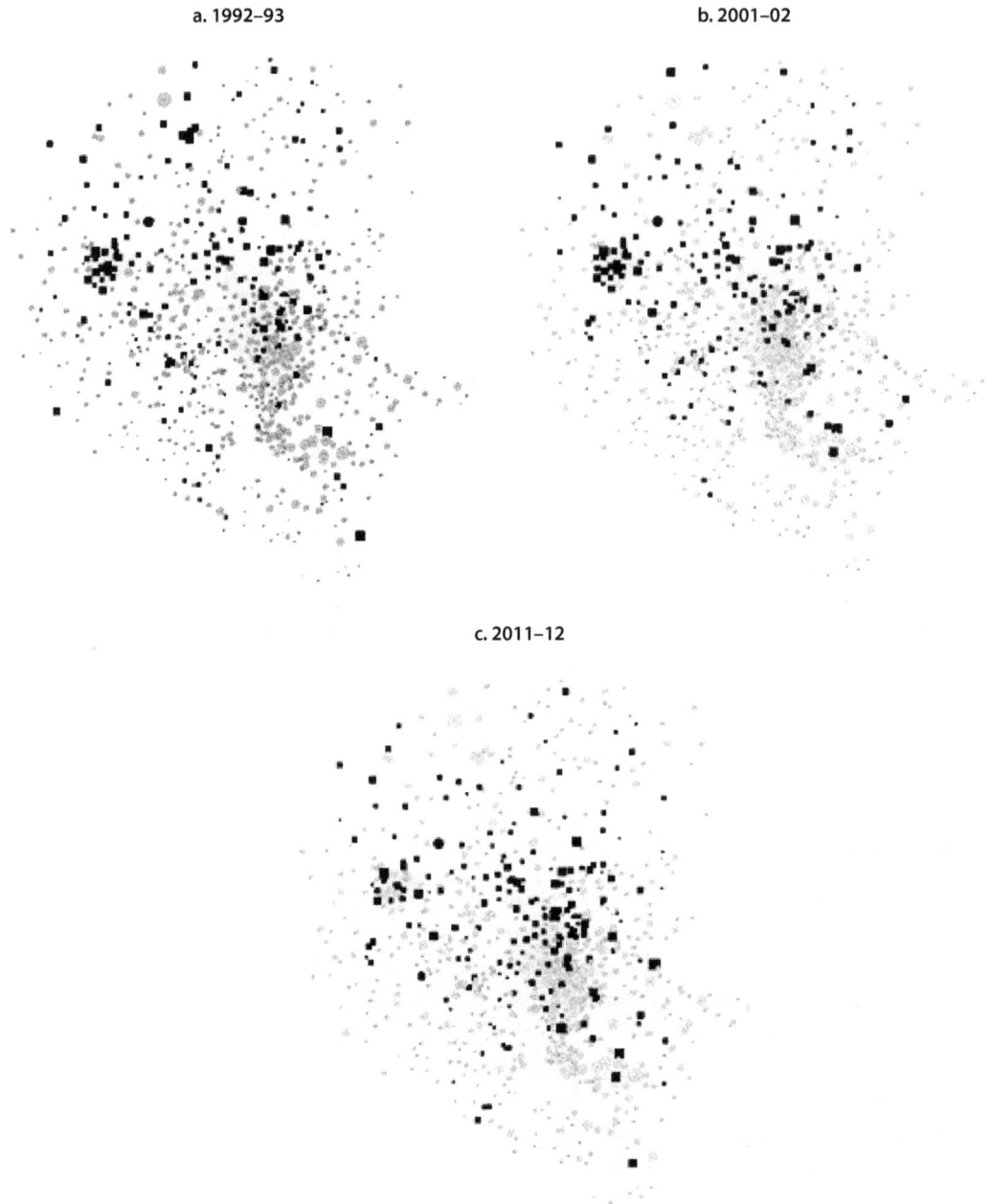

a. 1992–93

b. 2001–02

c. 2011–12

Note: Products with RCA>1 are marked by small black squares.

Table 3.4 Number of Exports with RCA>1

	1992–93	2001–02	2011–12
01. Food and beverages	36	32	39
02. Agricultural raw materials	14	14	17
03. Fuel, ores, metals	16	10	13
04. Chemicals	24	12	12
05. Material-based manufactures	51	52	47
06. Machinery and equipment	15	29	43
07. Apparel, footwear	21	25	13
08. Other manufactures	15	16	15
	192	190	199

Source: Calculated on the basis of data from UN Comtrade.

The latter are products in which Croatia has performed better (or attained more than its fair share) in world markets relative to the export performance of the world as a single economic unit.

Croatia has begun to have a foothold in advanced manufacturing activities. Figure 3.13 supports this view. Over the past decade in nominal terms, the apparel sector has had the slowest growth among all sectors, while chemical-related exports increased nearly threefold in value, although the number of exports decreased, suggesting greater consolidation around some successful products. The sector of machinery and equipment industries within manufacturing (SITC Section 7) was the best performing sector over the decade both for aggregate growth and for the number of products with enhanced competitiveness.

Identifying High-Potential Export Opportunities

Identifying new high-value exports begins with stock-taking the relative performance of existing exports. Table 3.5 filters products through the lens of significance in each of the four categories. We define "significant" exports as those with RCA>1 in both 2001–02 and 2011–12, of which there were 135 at the SITC 4-digit level, with 10 accounting for a quarter of the country's export earnings in 2012. The most significant export was that of ships and boats. We define "emerging" exports as those that had RCA<1 in 2001–02 but RCA>1 in 2011–12. There were at least 64 of these, including a set of four prominent products: medicaments (SITC 5417), specialized industrial machinery (SITC 7284), leather articles (SITC 6129), and engine parts (SITC 7149). These four emerging products alone accounted for 7 percent of merchandise earnings in 2011–12.

What is positive about this list is that more than half of the 12 top emerging exports are also highly complex in the knowledge and capabilities that they require. Indeed, specialized industrial machinery (SITC 7284) has been ranked as the world's most complex export to produce.[8] As seen in figure 3.14 below, many emerging exports are in the denser part of the product space, with far greater links to products with which they share productive knowledge.

Smart Specialization in Croatia • http://dx.doi.org/10.1596/978-1-4648-0458-8

Table 3.5 Evolution of Export Significance

Status	SITC Code	Product	RCA in 2002	RCA in 2012
Significant	7932	Ships and boats	22.0	9.6
Significant	7711	Electrical transformers	8.5	18.9
Significant	2483	Worked wood of non-coniferous	23.8	33.4
Significant	6842	Worked aluminum and aluminum alloys	3.7	4.9
Significant	5621	Nitrogenous fertilizers	9.8	10.9
Significant	7938	Special floating structures	7.8	7.4
Significant	8211	Chairs and seats	4.0	4.6
Significant	2820	Iron and steel waste	2.4	5.2
Significant	612	Refined sugar	7.9	14.2
Significant	7731	Electric wire	2.3	2.4
Significant	980	Edible products N.E.S.	4.8	3.9
Significant	8510	Footwear	3.8	2.2
Significant	5629	Fertilizers	11.8	8.2
Significant	6612	Cement	22.3	16.6
Emerging	5417	Medicaments	0.9	1.9
Emerging	7284	Specialized industry machinery and parts N.E.S	0.8	1.2
Emerging	6129	Other articles of leather	0.1	63.3
Emerging	7149	Parts of gas and reaction engines	0.6	2.6
Emerging	5530	Perfumery and cosmetics	0.3	1.4
Emerging	7499	Nonelectric parts of machinery N.E.S.	0.9	3.3
Emerging	3510	Electric wire	0.6	2.3
Emerging	7712	Parts of electric power machinery N.E.S.	0.9	1.5
Emerging	7491	Roller bearings	0.1	2.1
Emerging	7212	Harvesting and threshing machines	0.5	3.3
Emerging	5823	Polyesters	0.6	1.3
Emerging	11	Live bovines	0.2	5.7
Marginal	9710	Gold, nonmonetary	0.0	0.9
Marginal	7849	Other vehicles parts	0.4	0.5
Marginal	7721	Switchboards, relays, and fuses	0.7	1.0
Marginal	8939	Miscellaneous articles of plastic	0.7	0.7
Marginal	7492	Valves	0.6	0.7
Marginal	7139	Piston engines parts N.E.S.	0.7	0.8
Marginal	7763	Diodes and transistors	0.1	0.5
Declining	3414	Petroleum gases	2.0	0.7
Declining	7788	Other electrical machinery and equipment N.E.S.	1.4	0.3
Declining	8462	Knitted undergarments of cotton	3.4	1.0
Declining	5833	Polystyrene	4.0	0.7
Declining	8423	Men's trousers	3.0	0.5
Declining	8439	Other women's outerwear	1.9	0.2

Source: Calculated by authors on the basis of data from UN Comtrade.

Figure 3.14 Emerging Exports, 2012

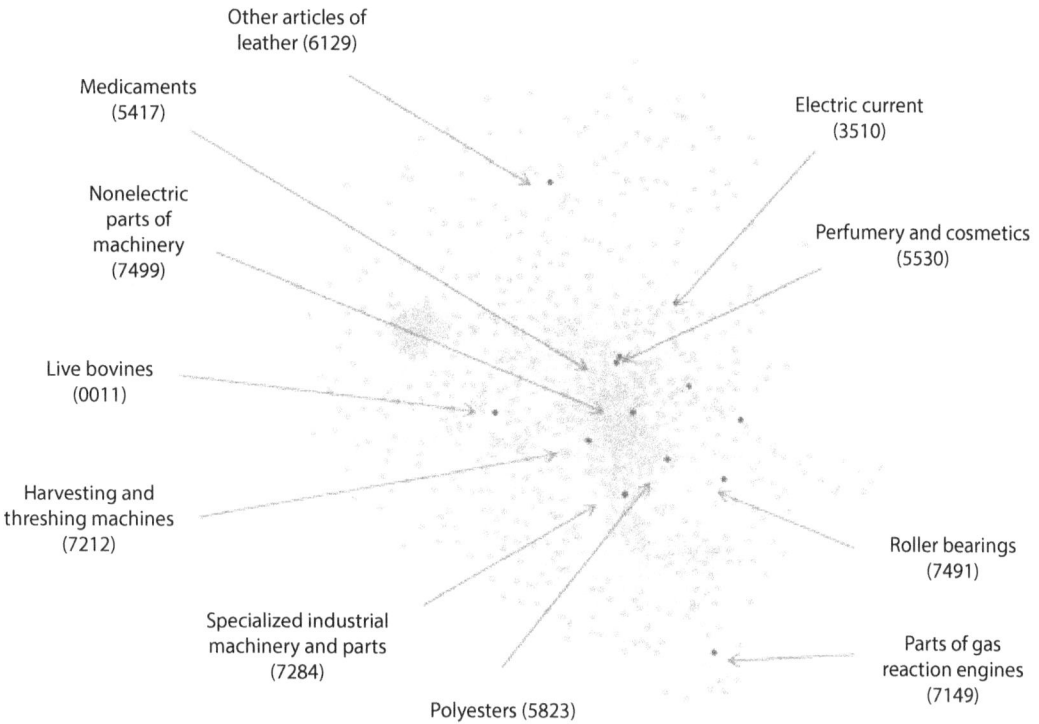

- Other articles of leather (6129)
- Medicaments (5417)
- Nonelectric parts of machinery (7499)
- Live bovines (0011)
- Harvesting and threshing machines (7212)
- Specialized industrial machinery and parts (7284)
- Polyesters (5823)
- Electric current (3510)
- Perfumery and cosmetics (5530)
- Roller bearings (7491)
- Parts of gas reaction engines (7149)

Hundreds of Croatia's exports are "marginal," but only 26 had a share of more than 0.1 percent in national exports in recent years. This threshold is not trivial, and it is an indication that these marginal exports at least are, potentially, not so marginal after all. Many are likely to evolve into emerging exports and possibly to become mainstays of Croatian trade, such as motor vehicle accessories (SITC 7849), electrical appliances (SITC 7721), parts of internal combustion engines (SITC 7139), taps, valves for industrial pipes (SITC 7139), and diodes and transistors (SITC 7763). These all require fairly complex production techniques and can potentially become part of global value chains (GVCs). Among notable marginals is an agricultural product, soya beans (SITC 2222), which contributed about 0.25 percent to total earnings in 2011–12. A deeper analysis of why these marginal products are still exported, yet have not expanded or become highly competitive, is necessary to improve the country's export promotion efforts. Interviews with local businesses, for example, revealed that some of the issues constraining the growth of small firms were the high cost of capital, labor regulations, red tape, and bureaucratic hurdles at country level; and not enough workers with applied skills. Finally, "declining" exports are those that had RCA>1 in 2001–02 but not in 2011–12. There were 55 of these, of which six used to be prominent with shares in 2001–02 greater than 0.75 percent.

Croatia seems set to do well in industrial and electronic manufacturing. Given what it exports at present and the capabilities it has built over the past decade, what is the basis for identifying the marginal products to be prioritized for scaling up? The distance between products on the map is used to compute how closely they are surrounded by the currently competitive basket of goods. "Higher density" implies the ease with which existing capabilities can be used to export products. Hausmann and Klinger (2007) showed countries to be more likely to export products with higher densities in the future because of their proximity to the current export basket. We calculate which of the products that we identified as marginal in the preceding analysis are closest to Croatia's existing export basket with RCA. Table 3.6 lists 26 marginal exports with a share of at least 0.1 percent; at least 22 of them belong to manufacturing, with an overwhelming majority belonging to machinery and transport equipment (SITC Section 7). This includes products as varied as lifting and loading machinery (SITC 7442),

Table 3.6 Prominent Marginal Exports with High Density

SITC code	Product	Density	Share in 2012	RCA 2012
1121	Wine	0.33	0.12	0.6
8124	Lighting fixture and lamp parts N.E.S.	0.32	0.10	0.5
5834	Polyvinyl chloride	0.32	0.12	0.8
6991	Base metal locksmiths wares N.E.S.	0.31	0.10	0.5
8939	Miscellaneous articles of plastic	0.31	0.40	0.7
7442	Lifting and loading machinery	0.30	0.20	0.7
7849	Other vehicles parts	0.30	1.09	0.5
7721	Switchboards, relays, and fuses	0.29	0.96	1.0
7492	Valves	0.29	0.32	0.7
7139	Piston engines parts N.E.S.	0.29	0.32	0.8
114	Poultry meat	0.28	0.11	0.7
7821	Trucks and vans	0.28	0.12	0.2
7452	Nonelectrical machines parts N.E.S.	0.27	0.23	0.8
7783	Auto parts	0.27	0.13	0.5
7649	Parts of telecom and sound recording equipment	0.27	0.10	0.1
7493	Mechanical tools for building	0.27	0.12	0.4
6940	Nails, nuts, and bolts	0.27	0.18	0.8
7781	Batteries	0.26	0.11	0.5
6954	Interchangeable hand and machine tools	0.26	0.12	0.6
9710	Gold, nonmonetary	0.26	1.16	0.9
7415	Air conditioning machines	0.26	0.11	0.5
7234	Construction and mining machinery	0.26	0.16	0.4
5989	Chemical products	0.23	0.22	0.3
2222	Soya beans	0.23	0.24	0.8
7929	Aircraft equipment parts N.E.S.	0.21	0.13	0.4
7763	Diodes and transistors	0.20	0.30	0.5

Source: Calculated by authors on the basis of data from UN Comtrade.

switchboards, relays, and fuses (SITC 7721), piston engine parts (SITC 7139), auto parts (SITC 7783), and mechanical tools for building (SITC 7493).

Croatia also has an advantage in processed food and beverages. The list of high-potential marginal exports also includes processed food and beverages such as wine (SITC 1121) and poultry meat (SITC 0114). In many trade classifications, processed food and beverages are subsumed in the broader category of "primary" (agricultural goods). Yet processed foods have highly desirable properties (Athukorala and Waglé 2013): first, the income and price elasticities of demand for such foods are higher than those for most traditional primary agricultural products, and so diversification of the export mix into this commodity category can bring faster export growth and terms-of-trade gains. Second, the final stages of food processing are labor intensive, unlike minerals or other material-based manufacturing, helping create jobs. And third, processed food products typically have greater domestic input content and so greater potential for domestic value addition.

Building on the history of food as a strategic sector during the Yugoslav years, Croatia is well placed geographically to become a pan-European food hub. As the McKinsey Global Institute (2013) pointed out, average Central European labor costs are about one-quarter of those in Western Europe; savings in materials and other costs outweigh the higher transport costs. But to become such a hub, Croatia would require new investments to consolidate and modernize contract farming, and to provide know-how and capital for state-of-the-art techniques. Indeed, Croatia's largest private company and regional multinational, Agrokor, is part of this industry, with 40,000 employees and over €4 billion revenue. It has developed an integrated farm-to-retail supply chain in food processing, after years of in-house innovation, large investments in farms, and tie-ups with research houses abroad for specific technologies.[9]

Many emerging and marginal exports are embedded in dense product networks, which augurs well for future growth. Several emerging and marginal products are close to the more desirable clusters of the product space (figures 3.14 and 3.15), where networks are dense. Their embeddedness in dense clusters suggests that it would be easy for the country to undergo structural transformation because of the higher degree of shared knowledge of production techniques.

Significant exports, in contrast (figure 3.16), are in the less dense parts of the product space. This suggests that the capabilities and knowledge entailed by some of Croatia's most significant exports do not lend themselves readily to redeployment. In other words, capabilities built by, for example, firms engaged in sawing wood or producing cement may be so specific that they are difficult to reapply to other industries.

Some formerly important exports show declining significance. The most conspicuous declining or stagnating exports are related to garments (SITC 8462, 8423, and 8439; figure 3.17). These are exports that had RCA>1 in 2001–02 but that no longer had such an RCA value in 2011–12. It is unclear to what extent these products lost out to cheaper competition from Asia in the aftermath of the

Figure 3.15 Marginal Exports, 2012

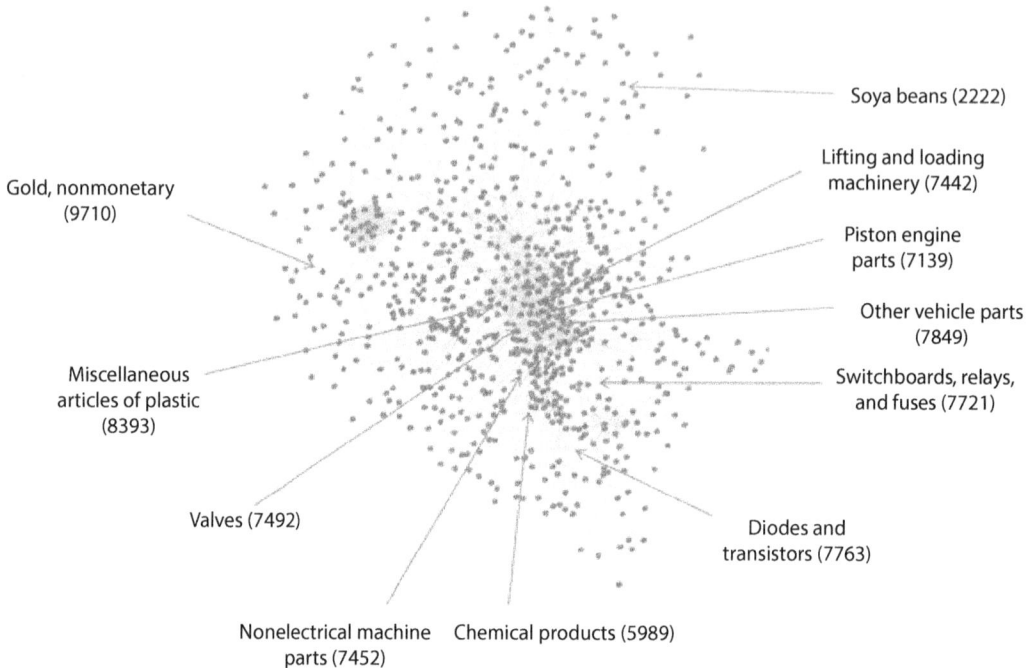

abolition of quotas that restrained the export of textiles and clothing from developing countries before 2005.[10] Globally in the 2000s, there was consolidation of labor-intensive and price-sensitive segments of the industry toward Asian countries led by China. Croatia still has the advantage of proximity to the European market in higher-value, and perhaps fashion-sensitive, apparel requiring short lead times. Indeed, while the export values of products like knitted cotton undergarments have dropped by half over the past decade, they still earned US$25 million in 2012. Further, Croatia appears to be moving toward specializing in high-value, niche items such as performance textiles. The top six declining exports saw their contribution to total earnings drop from about 5 percent in 2001–02 to 1.6 percent in 2011–12. The only export, also deemed relatively complex, that appears to be in relative decline is polystyrene (SITC 5833).

"Low-Hanging Fruits" for Croatia Are Few and Far Between

Which marginal exports hold more promise of growing into bigger exports? In identifying potential new exports, one should consider global demand as well as the inherent quality or complexity of the product. All else being equal, products with high import growth would be more desirable than products that are stagnant or declining in global markets. Of the 532 marginal Croatian exports, we analyze the collective prospects of about 160 of

Figure 3.16 Significant Exports, 2012

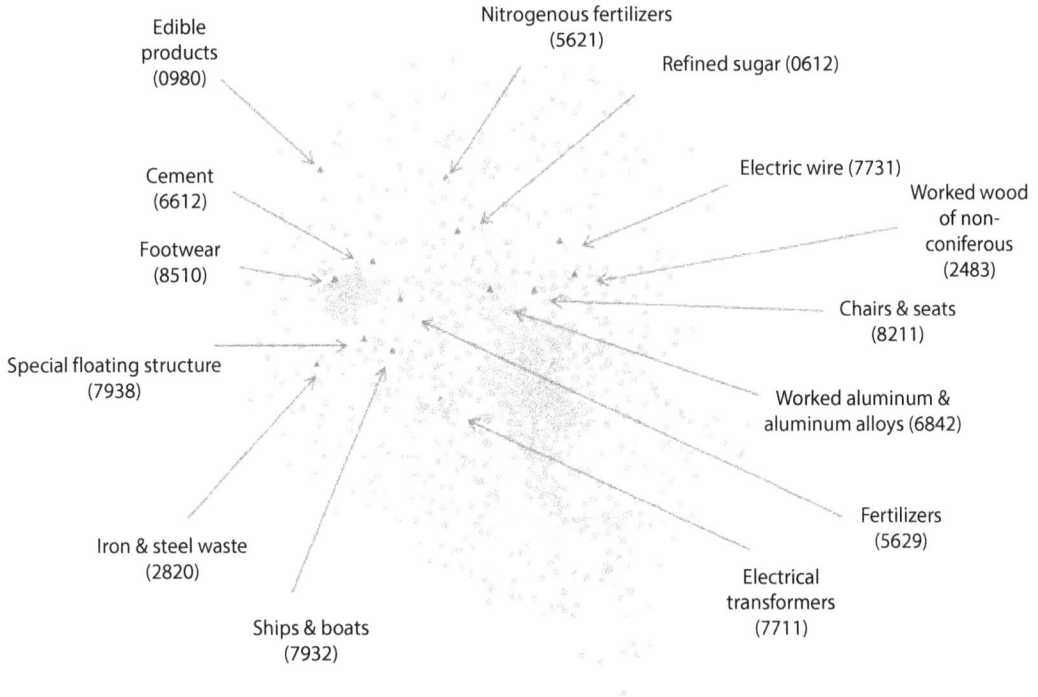

Edible products (0980)
Nitrogenous fertilizers (5621)
Refined sugar (0612)
Electric wire (7731)
Worked wood of non-coniferous (2483)
Cement (6612)
Footwear (8510)
Chairs & seats (8211)
Special floating structure (7938)
Worked aluminum & aluminum alloys (6842)
Fertilizers (5629)
Iron & steel waste (2820)
Electrical transformers (7711)
Ships & boats (7932)

Figure 3.17 Declining Exports, 2012

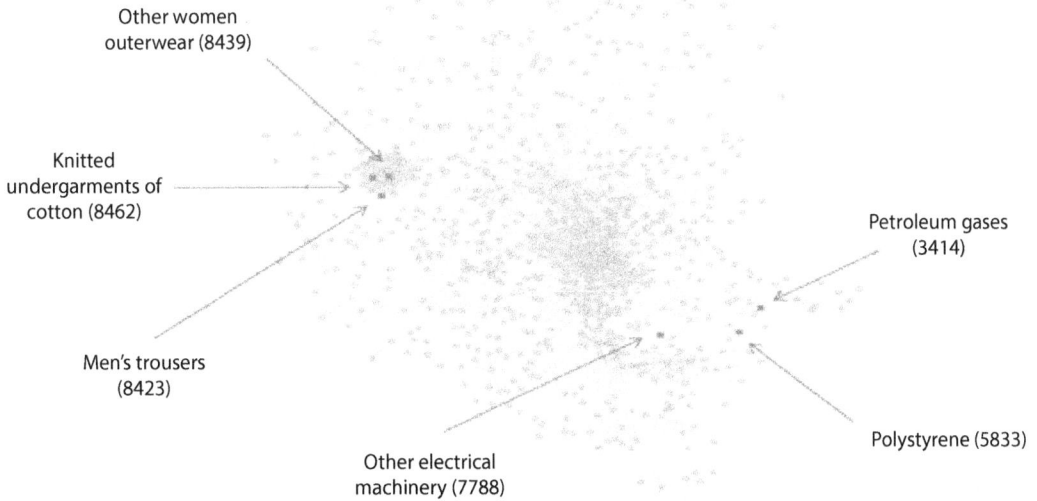

Other women outerwear (8439)
Knitted undergarments of cotton (8462)
Petroleum gases (3414)
Men's trousers (8423)
Polystyrene (5833)
Other electrical machinery (7788)

them—those with at least 0.01 percent share in national exports in 2011–12. Figure 3.18 identifies Croatia's marginal exports that are in the top quintile (20 percent) for global import growth and for worldwide import share in 2002–12: there are 17 such products. Six are quite close to the existing basket of competitive goods (within and below one standard deviation of the mean density, indicated by two dashed lines).

Of the six products, five are manufactured goods related to the resources sector (materials-based manufactures belonging to SITC Section 6). They had healthy growth over the past decade because of high demand in emerging markets. The five are iron/steel rods (SITC 6732), ferro-alloys (SITC 6716), materials of rubber (SITC 6210), worked copper and copper alloys (SITC 6822), and unwrought silver (SITC 6811). Among products that are less proximate (to the right of the second dashed line) yet still significant globally and quite large in Croatian exports are nonmonetary gold (SITC 9710), construction and mining machinery (SITC 7234), digital data-processing machines (SITC 7522), optical instruments (SITC 8710), and diodes and transistors (SITC 7763). This enumeration reaffirms the previous observation about the trade-off between complexity and proximity, because the more sophisticated manufactured exports belonging to SITC Section 1 (machinery and transport equipment) are further from the capabilities embodied in the existing basket of competitive exports.

Other marginal exports that are highly proximate (less than two standard deviations from the mean) and that grew globally at above-average rates over the past decade include the following, but they do not rank high on product complexity: iron or steel tubes and pipes (SITC 6783), containers for transportation (SITC 7861), centrifugal pumps parts (SITC 7449), and other textile

Figure 3.18 Proximity of Croatian Exports to High Global Demand

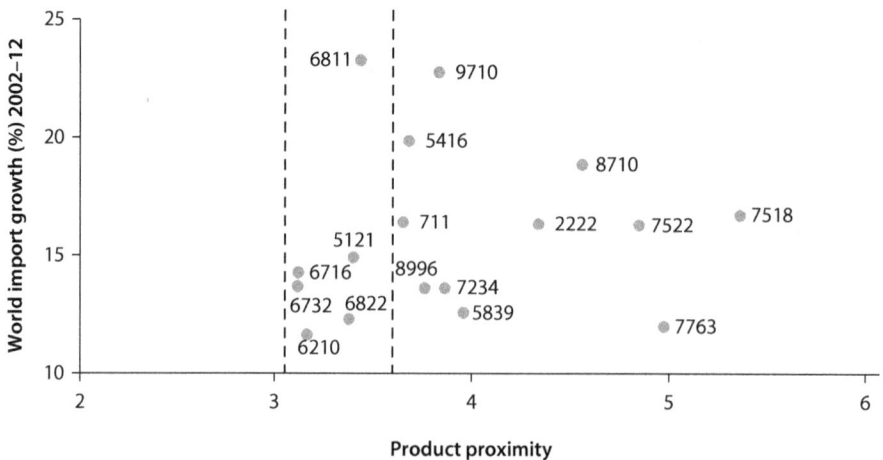

Source: Authors' calculation based on UN Comtrade data.
Note: Any products to the left of the first dashed line are the most proximate and have more productive knowledge in common with the country's basket of competitive goods. Products with densities above the mean value of 3.6 on the x-axis are less proximate.

articles (SITC 6589). Among the proximate marginal exports (less than one standard deviation from the mean) that grew at above-average rates globally and are commercially significant in world trade, the following 10 products, all in manufacturing, stand out: switchboards, relays, and fuses (SITC 7721); valves (SITC 7492); lifting and loading machinery (SITC 7442); other worked iron or steel sheets (SITC 6749); liquid and gas filters and purifiers (SITC 7436); acyclic alcohols and derivatives (SITC 5121); polypropylene (SITC 5832); iron/steel rods (SITC 6732); color television (SITC 7611); and worked copper and copper alloys (SITC 6822). In addition to the products already mentioned (such as color TVs, valves, and lifting/loading machinery), proximate products that have gained market share include acyclic alcohols and derivatives (SITC 5121), iron/steel rods (SITC 6732), and other worked iron/steel sheets (SITC 6749).

Finally, among Croatian exports that have some of the desirable traits mentioned above, yet are far from the competencies acquired by the existing export basket, are the following: aircraft equipment parts N.E.S. (SITC 7929), chemical products (SITC 5989), heterocyclic compound nucleic acids (SITC 5156), drawing and mathematical calculating instruments (SITC 8742), pro-vitamins and vitamins (SITC 5411), and scientific instruments N.E.S. (SITC 8745). All these products are highly complex and share many of the desirable properties for growth and share of world trade. They are produced in Croatia in small quantities and could be candidates for scaling up.

Could Croatia Become a Bigger Player in GVCs?

Croatia's nascent positioning in GVCs, on products and markets, is solid, but the country needs to do much more to maximize this potential. How then does it currently perform? One synthetic way of answering this question is through a network representation of trade in intermediates, which helps one grasp the increasing interdependence of countries worldwide, although visualizing large and dense networks may be daunting given the many links to be drawn. One way to overcome the problem is to use a tree representation. A "tree is a connected, undirected network with no closed loops" (Newman 2010).

The minimal spanning tree in figure 3.19 depicts where Croatia stood on a global network of bilateral trade in intermediate goods in 2010. By trees drawn using a rooted algorithm, link weights have been transformed to reflect distances between trade nodes. The EU—in particular, Italy—is the most relevant trading partner for Croatia (circled in red). Croatia is rather marginal to the global focal nodes of trade in intermediates—China, Germany, and the United States—but at least for intermediate goods from Bosnia and Herzegovina it is the most important trade partner. A closer look at a few key sectors sheds further light on the potential for Croatia to develop along GVCs.

Croatia missed the rush of foreign direct investment (FDI) in transport vehicles, parts, and equipment in the 1990s and early 2000s. This product group (SITC Divisions 78 and 79) is one of the most dynamic in world trade, partly

Figure 3.19 Minimal Spanning Tree, Exports of Intermediates, 2010

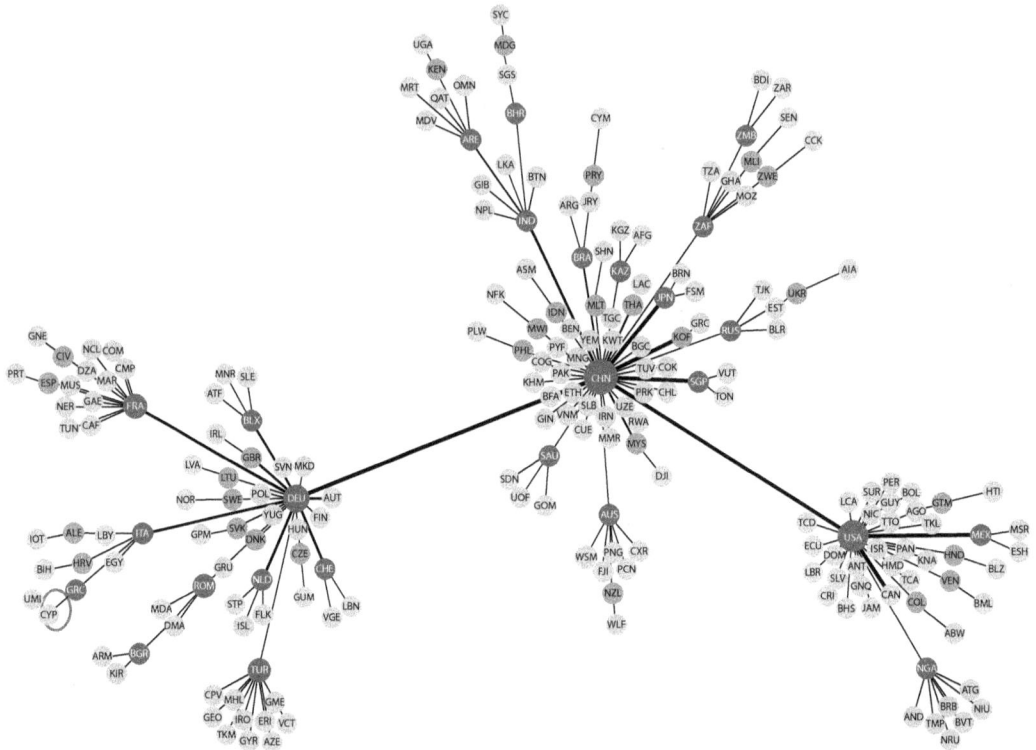

Note: Larger bilateral trade flows are portrayed by closer distances between nodes. The result is an acyclic subset of edges that connects all the nodes in the network such that the total weight is minimized—the so-called minimal spanning tree. The most connected countries are the roots of the tree. Such hubs, which represent the main trading partner for several countries, are drawn using darker colors and bigger shapes. Peripheral countries, so-called leaves, are in lighter colors. Link weights are proportional to trade flows, while the size of the nodes reflects a country's degree of centrality.

because of its fragmentability of production across borders. At present, unlike several countries in Central Europe that have attracted western European and Asian FDI into new plants or whose old factories from the socialist era were acquired, Croatia has no auto-manufacturing cluster. Among notable sales were Volkswagen's acquisition of Skoda in the Czech Republic in 1991; Fiat's purchase of Poland's FSM in 1992; and Renault's purchase of Dacia in Romania in 1998. Audi, Opel, and Suzuki have opened new plants in Hungary; Peugeot, Toyota, and Hyundai have operations in the Czech Republic; and PSA Peugeot Citroen has investments in the Slovak Republic. Because of the heavy reliance on just-in-time production and the high weight-to-value ratio of auto parts and components, their suppliers tend to locate close to auto manufacturers, helping to create clusters. Croatia was sidelined by the FDI boom a decade or so ago both because of the war and because it did not have a strong auto base before the 1990s to make it attractive for brownfield investment.

Croatia has emerging success in several industrial machinery and electronic products. The nature of procurement of parts and components within global

Figure 3.20 Exports of Parts and Components, Croatia and Six Regional Peers

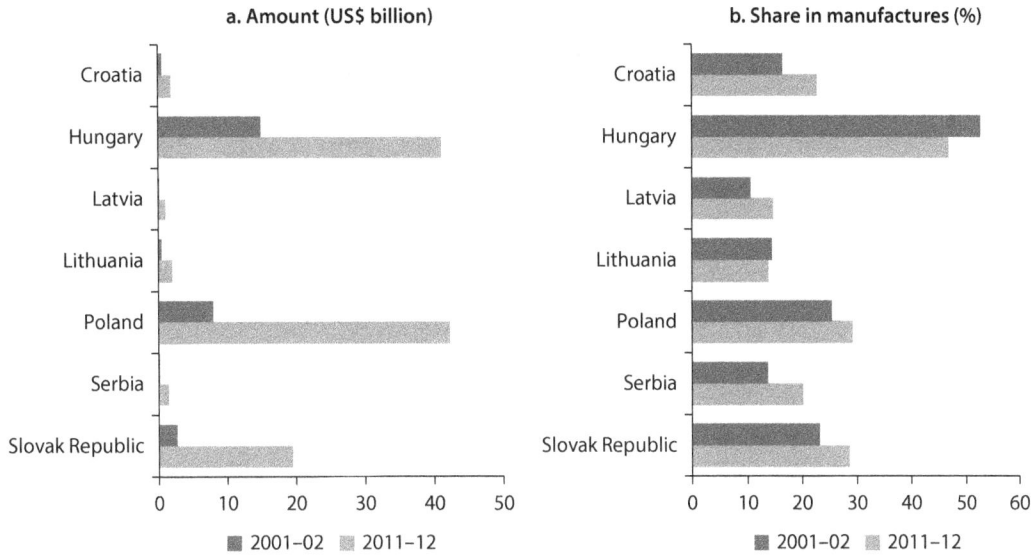

a. Amount (US$ billion)

b. Share in manufactures (%)

■ 2001–02 ▨ 2011–12

Source: UN Comtrade.
Note: Manufactures cover SITC sections 5–8, net of division 68.

production networks differs between the auto industry and other manufacturing industries. Sourcing of parts across borders (or longer distances) is common in the electrical and electronics industry, for example. While it is unclear to what extent Croatia is already part of GVCs in the non-auto industries, the export of parts and components has more than tripled in the past decade, from about US$0.6 billion to over US$2 billion. This marks respectable nominal growth (13 percent a year), but is still lower than that among regional peers—Latvia (22.4 percent), Lithuania (17 percent), Poland (18 percent), Serbia (24 percent), and the Slovak Republic (22 percent). But its share of parts and component exports in overall manufactured exports is higher than in some peers (figure 3.20).

Most of Croatia's exports of parts and components go to Europe (EU-27 and CEFTA[11]), its natural destination given the proximity. In 2011–12, Europe accounted for about 60 percent of overall merchandise exports, and nearly 20 percent went to CEFTA countries; for parts and components, the corresponding shares were 78 percent and 7.5 percent. Within Europe, the top destinations for parts and components in 2011–12 were Germany (US$430 million), Austria (US$303 million), Slovenia (US$219 million), Italy (US$151 million), and Bosnia and Herzegovina (US$66 million).

The Importance of Services Exports

A distinguishing feature of Croatia's economy is the importance of services exports. While similar countries export a greater value of goods than the value of services, Croatia's services exports are about equal to its goods exports (table 3.7).

Table 3.7 Exports of Goods and Services, Croatia and Three Regional Peers, 2012
US$ billion

Country	Goods	Services	Services/goods ratio (%)
Croatia	12.3	12.0	97
Slovak Republic	81.2	7.0	9
Slovenia	32.1	6.4	20
Serbia	11.4	4.0	35

Figure 3.21 Croatia vs. Peers, per Capita Services Exports vs. Income, 2002–04 vs. 2008–10

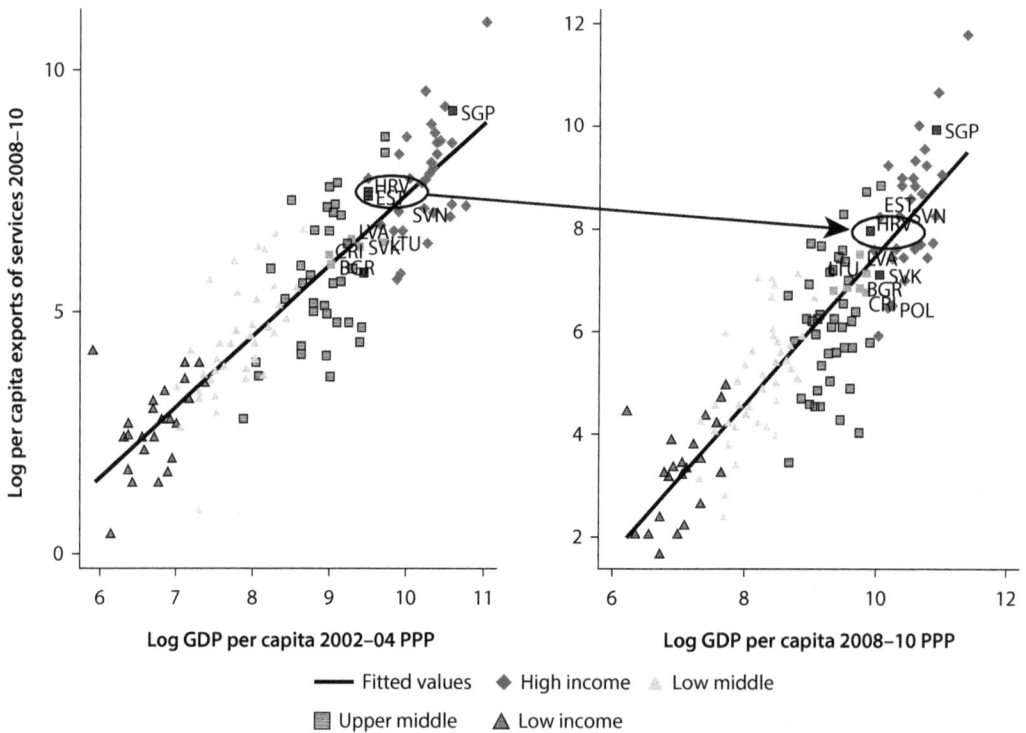

Note: The upward-sloping line indicates the fitted values obtained by a linear prediction of the relationship between the two variables. It indicates that per capita services exports are larger in richer countries.

Croatia's per capita services exports are higher than predicted on the basis of the country's income (figure 3.21), which shows them above the regression line. Across the peer countries, only Estonia shows clearly higher per capita exports relative to its income, while other peers, notably Poland, lag behind.

Croatia's services exports are, though, unusually concentrated in tourism, whereas high-income countries' services exports are more heavily concentrated in business services. With its share expanding from 68 percent in 2000 to

73 percent in 2012, travel dominates commercial services exports. Travel services exports grew an average of 10 percent a year in 2000–12, as the number of international arrivals climbed from around 5.8 million to 9.9 million. (Travel services exports grew even faster, at 19 percent a year in 2000–08, i.e., before the global crisis.) The fact that the share of travel still expanded implies that other subsectors grew at slower rates. The large share of travel reflects its RCA growth, which rose strongly over 2000–09 to become the highest across all services subsectors (figure 3.22).

Transport and other business services together make up one-fifth of commercial services exports. They had similar shares of around 10 percent in 2012, marking a decline in the share of transport services exports from 13.7 percent to 10.1 percent over 2000–12 (reflecting a strong decline in its RCA after 2000 to below 1), while other business services expanded their share from 6.5 percent to 9.8 percent (although they also had RCA below 1).

The lack of dynamism of Croatia's services is reflected in the country's stagnating value added of services over recent years, which saw instead increasing "servicification" of manufacturing worldwide. While services capture more than two-thirds of Croatia's value added—68 percent in 2010—this share expanded only slightly over the past decade, in contrast to the strong growth from 53 percent to 65 percent in 1990–2000 (figure 3.23). Yet the country's share of services in value added is higher than that in other countries of similar income levels (figure 3.24). To better capture the more complex relationship between value added in services and GDP per capita,

Figure 3.22 RCA by Services Subsector, Croatia, 2000, 2005, and 2009

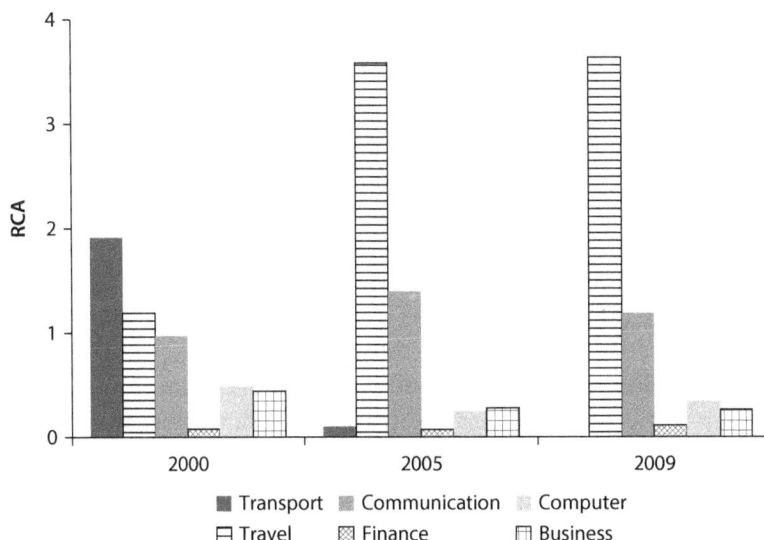

Source: Authors' elaboration based on Saez et al. (2012).

Figure 3.23 Croatia vs. Peer Countries, Services Sector Share in the Value Added, 1990, 2000, and 2010 Sector Share in Value Added vs. GDP per Capita, 2008–10

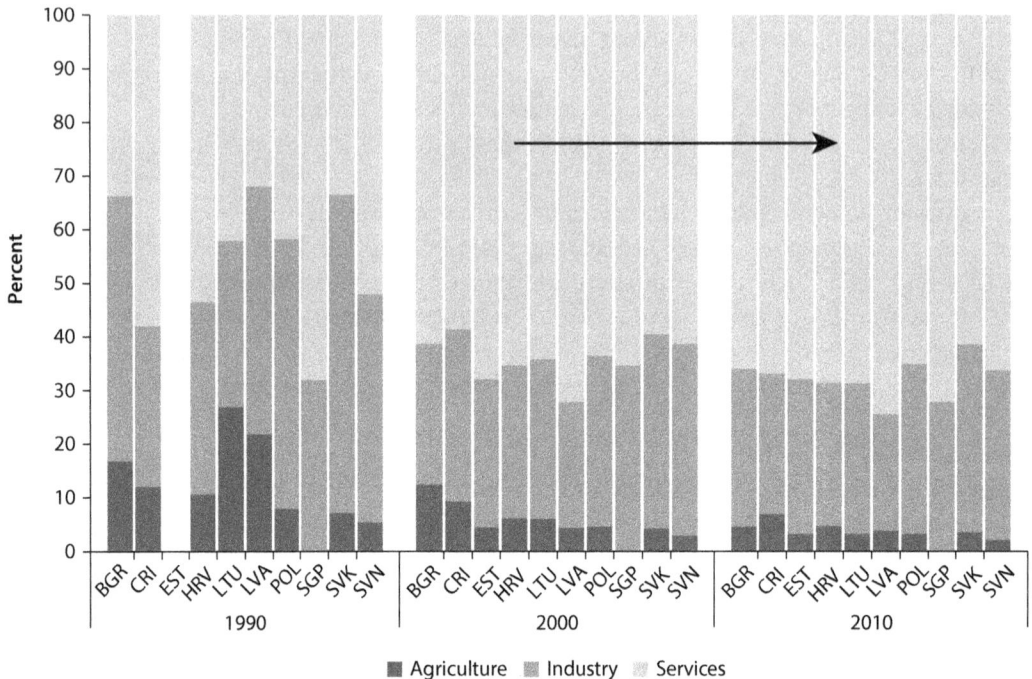

Agriculture   Industry   Services

Source: Authors' illustration based on World Development Indicators.
Note: In the World Development Indicators database, services correspond to ISIC divisions 50–99. This sector is derived as a residual (from GDP less agriculture and industry) and may not properly reflect the sum of services output, including banking and financial services. For some countries, it includes product taxes (minus subsidies) and may also include statistical discrepancies.

a linear regression including a quadratic GDP term was applied. The figure also includes the 95 percent confidence intervals around the fitted line to show to what extent a country is an outlier relative to the prediction. The upward-sloping line indicates the fitted values obtained by a linear prediction of the relationship between the two variables. The upward slope reflects the stylized fact that the services sector represents a larger share of the economy for richer countries. Croatia's share of services in value added lies slightly above the fitted line (outside the 95 percent confidence intervals), suggesting (again) that it exceeds what would be predicted based on its income.

Croatia's services export sophistication is behind that of its peers (figure 3.25). Despite services exports showing greater sophistication, they were the lowest over the period and appear to be falling farther behind. This may be related to the share of tourism increasing and that of more sophisticated services, such as transport and other commercial services, declining.

Figure 3.24 Sector Share in Value Added vs. GDP per Capita, 2008–10

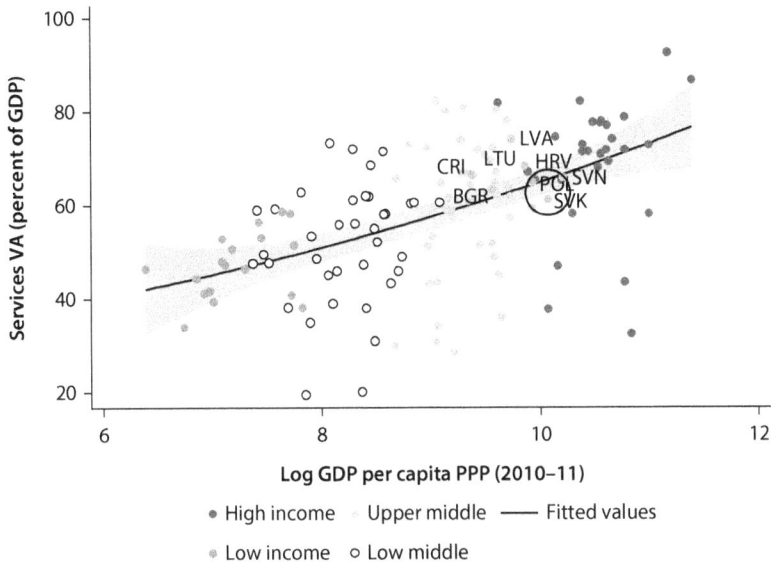

Source: Authors' illustration based on World Development Indicators.
Note: In the World Development Indicators database, services correspond to ISIC divisions 50–99. This sector is derived as a residual (from GDP less agriculture and industry) and may not properly reflect the sum of services output, including banking and financial services. For some countries, it includes product taxes (minus subsidies) and may also include statistical discrepancies.

Figure 3.25 Services Export Sophistication, Croatia vs. Peers, 2000–09

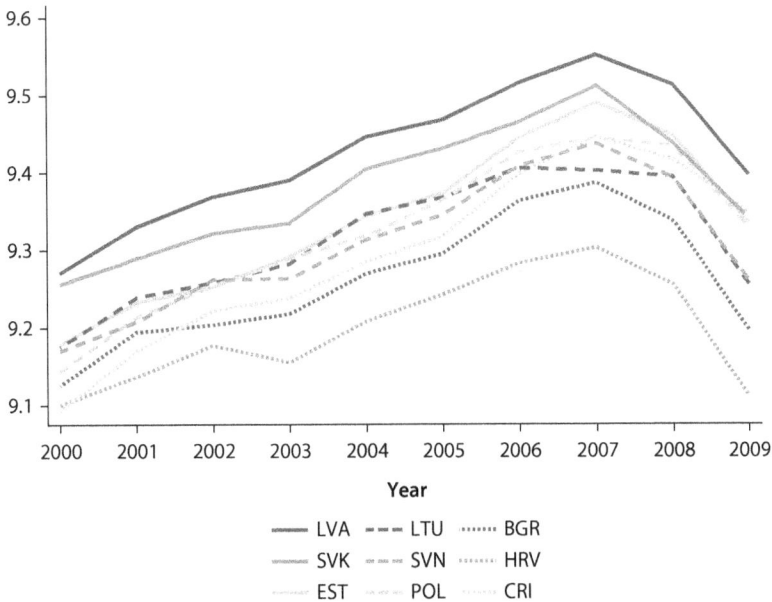

Source: Authors' elaboration based on IMF Balance of Payments Statistics database.

Notes

1. The authors of this chapter are: Michael Ferrantino, Daria Taglioni, and Swarmim Waglé. Additional input on specific issues was provided by Deborah Winkler, Guillermo Arenas, Claire Hollweg, and Jose Daniel Reyes.

2. Some analysis of this section is carried out by SITC classification and not HS classification. Products classified according to HS at the 6-digit level are the most disaggregated available for international comparison. More than 5,000 products are classified this way, and they are our preferred choice of trade nomenclature for the most precise description and analysis possible. In contrast, SITC at the 4-digit level classifies between around 780 (in Revision 2) and 1,033 products (Revision 3)—a more aggregated classification. The stage of processing in this SITC classification is easier to identify than that in an HS classification, as the latter emphasizes the description of the material properties of goods. In most cases they reflect finely grouped industries rather than individual products. The use of SITC is preferred when the analysis requires discussion of industrial transformation and when there is a need for a long data series: HS begins in 1988, whereas SITC (Revision 2) begins in 1961. Further, indexes of product and economic complexity used for this report have been computed using only data classified according to SITC at the 4-digit level (Revision 2).

3. As mentioned, when the greatest possible disaggregation is desirable such as the analysis of quality upgrading at the level of individual exports, we prefer to analyze products at the HS 6-digit level, with over 5,000 merchandise products.

4. The indexes are computed by weighting the factor endowments of all countries exporting a particular product; weights are derived from a modified RCA. Human capital is estimated by the average years of schooling, and physical capital stock is estimated by the perpetual inventory method, which reconstructs capital stock estimates from investment flows by recursively adding up current investments to the previous period's capital stock with appropriate depreciation. Goods that are predominantly exported by countries rich in human (physical) capital are revealed to be intensive in human (physical) capital (Shirotori, Tumurchudur, and Cadot 2010).

5. This includes petroleum gas (HS 271000) and tankers (HS 890120).

6. Quality-elastic industries are those where the positive difference between export and import unit values leads to a trade surplus, that is, a positive difference between export and import quantity; low relative unit value, in contrast, leads to a trade deficit.

7. We only analyze exports with non-missing values in 2000 and 2010.

8. Based on averaged product complexity indexes for 2007–09, computed by the Observatory of Economic Complexity.

9. Similar observations were made by a senior executive in Atlantic Grupa, another food company with a strong presence in the former Yugoslav states.

10. Between 1974 and 1994, exports of textiles and clothing were governed by quotas negotiated through the Multi-Fiber Arrangement. The successor Agreement on Textiles and Clothing ended all quotas on December 31, 2004, after which countries have not been restricted quantitatively.

11. European Free Trade Association

Bibliography

Aiginger, K. 2001. "Measuring the Intensity of Quality Competition in Industries." *Austrian Economic Quarterly* 6 (2): 73–101.

Anos-Casero, Paloma, and Charles Udomsaph. 2009. "What Drives Firm-Level Productivity?" Policy Research Working Paper 4841, World Bank, Washington, DC.

Athukorala, P., and S. Waglé. 2013. "Export Performance in Transition: The Case of Georgia." Working Papers in Trade and Development 2, Australian National University, Canberra.

Charron, N., V. Lapuente, and L. Dykstra. 2012. "Regional Governance Matters: A Study on Regional Variation in Quality of Government within the EU." Regional Studies Working Paper 1.

Cowey, Lisa (ECORYS). 2013. "Analysis of S&T Strengths and Strategically Supporting Soft and Hard Innovation Infrastructure Developments." Presentation, November 25.

Dunning, John H., and Sarianna M. Lundan. 2008. *Multinational Enterprises and the Global Economy*. 2nd ed. Cheltenham, UK: Edward Elgar.

Ericsson Nikola Tesla d.d. *Annual Report 2012*. http://www.ericsson.hr/homepage.

Europe Innova/Pro Inno Europe. 2009. "The Concept of Clusters and Cluster Policies and Their Role for Competitiveness and Innovation: Main Statistical Results and Lessons Learned." European Commission: Europe Innova/Pro Inno Europe Paper 9, Brussels.

European Commission. 2013. *Innovation Union Scoreboard 2013*. Brussels: European Commission.

Ferrantino, M. J., R. Feinberg, and L. Deason. 2012. "Quality Competition and Pricing-to-Market: A Unified Framework for the Analysis of Bilateral Unit Values." *Southern Economic Journal* 78 (3): 860–77.

Francois, J., M. Manchin, & P. Tomberger. 2013. "Services Linkages and the Value Added Content of Trade." Policy Research Working Paper Series 6432, World Bank.

Gaulier, G., G. Santoni, D. Taglioni, and S. Zignago. 2013. "In the Wake of the Global Crisis: Evidence from a New Quarterly Database of Export Competitiveness." Policy Research Working Paper 6733, World Bank, Washington, DC.

Handjiski, B., R. Lucas, P. Martin, and S. S. Guerin. 2010. "Enhancing Regional Trade Integration in Southeast Europe." Working Paper 185, World Bank, Washington, DC.

Hausmann, R., C. A. Hidalgo, S. Bustos, M. Coscia, A. Simoes, and M. A. Yildirim. 2011. *The Atlas of Economic Complexity: Mapping Paths to Prosperity*. Hollis, NH: Puritan Press.

Hausmann, R., and B. Klinger. 2007. "The Structure of the Product Space and the Evolution of Comparative Advantage." CID Working Paper 146, Harvard University, Cambridge, MA.

Hausmann, R., D. Rodrik, and J. Hwang. 2007. "What You Export Matters." *Journal of Economic Growth* 12 (1): 1–25.

Helpman, E., M. Melitz, and Y. Rubinstein. 2008. "Estimating Trade Flows: Trading Partners and Trading Volumes." *Quarterly Journal of Economics* 123 (2): 441–87.

Hidalgo, C. A., B. Klinger, A.-L. Barabasi, and R. Hausmann. 2007. "The Product Space and Its Consequences for Economic Growth." *Science* 317: 482–87.

Hummels, David, and Peter L. Klenow. 2005. "The Variety and Quality of a Nation's Exports." *American Economic Review* 95 (3): 704–23.

Hwang, J. 2006. "Introduction of New Goods, Convergence and Growth." Job Market Paper, Department of Economics, Harvard University, Cambridge, MA.

IMF (International Monetary Fund). "Balance of Payments Statistics database."

Kommerskollegium (Swedish National Board of Trade). 2010. "At Your Service: The Importance of Services for Manufacturing Companies and Possible Trade Policy Implications." Kommerskollegium, Stockholm.

Koopman, R., Z. Wang, and Shang-Jin Wei. 2012. "Estimating Domestic Content in Exports When Processing Trade Is Pervasive." *Journal of Development Economics* 99 (1): 178–89.

Krugman, P. 1991. *Geography and Trade*. Cambridge, MA: MIT Press.

Lederman, D., and W. Maloney. 2012. "Does What You Export Matter? In Search for Empirical Guidance for Industrial Policies." World Bank, Washington, DC.

Mahadevan, R. 2007. "The Poverty Transition: When, How and What Next?" *Journal of International Development* 19 (8): 1099–113.

Marshall, A. 1920. *Principles of Economics*. 9th ed. London: Macmillan.

McKinsey Global Institute. 2013. *A New Dawn: Reigniting Growth in Central and Eastern Europe*.

Medina-Smith, E. 2001. "Is the Export-Led Growth Hypothesis Valid for Developing Countries? A Case Study of Costa Rica." Policy Issues in International Trade and Commodities Study Series 7, United Nations Conference on Trade and Development, Geneva, Switzerland.

Mudabi, R. 2008. "Location, Control and Innovation in Knowledge-Intensive Industries." *Journal of Economic Geography* 8 (5): 699–725.

Newman, Mark E. J. 2010. *Networks: An Introduction*. Oxford, UK: Oxford University Press.

Powell, Walter W., and Stine Grodhal. 2005. "Networks of Innovators." In *The Oxford Handbook of Innovation*, edited by Jan Fagerberg, David C. Mowery, and Richard R. Nelson, 56–85. Oxford, UK: Oxford University Press.

Rabellotti, R., A. Carabelli, and G. Hirsch. 2009. "Italian Industrial Districts on the Move: Where Are They Going?" *European Planning Studies* 17 (1): 19–41.

Racine, J., ed. 2011. "Harnessing Quality for Global Competitiveness in Eastern Europe and Central Asia." World Bank, Washington, DC.

Reis, J., and T. Farole. 2012. "Trade Competitiveness Diagnostic Toolkit." World Bank, Washington, DC.

Rodríguez-Pose, Andrés, and Tobias Ketterer. 2013. "The Determinants of Regional Growth in the EU and their Implications for Development Strategies in West Romania." Background paper for World Bank Romania West Region: Competitiveness Enhancement and Smart Specialization Strategy, Final Report.

Schott, P. 2004. "Across Product versus Within Product Specialization in International Trade." *Quarterly Journal of Economics* 119 (2): 646–77.

Shirotori, M., B. Tumurchudur, and O. Cadot. 2010. "Revealed Factor Indices at the Product Level." Policy Issues in International Trade and Commodities Study Series 44, United Nations Conference on Trade and Development, Geneva, Switzerland.

UN Comtrade database. http://comtrade.un.org.

UNCTAD database. http://unctad.org/en/pages/Statistics.aspx.

Wagner, Joachin. 2007. "Exports and Productivity: A Survey of the Evidence from Firm-Level Data." *The World Economy* 30 (1): 60–82.

WBC-INCO.NET. 2011. "Innovation Infrastructures: Croatia." Coordination of Research Policies with the Western Balkan Countries.

CHAPTER 4

Stylized Facts on Productivity

Chapter Summary[1]

Four conclusions stand out from this chapter's review of productivity in a sample of Croatian firms. First, heterogeneity of firm performance in Croatia increased over 2008–12 on labor productivity, capital productivity, and unit labor costs. (Knowledge of heterogeneity is important for policy makers to prioritize investments or explore complementarities among firm characteristics.) On firm heterogeneity across different analytical criteria—region, size, export orientation, and ownership—controlled for sector-level differences, the following findings emerged: Adriatic region firms fare worse that Continental region firms; trade-integrated firms have better performance than partly or nontrade-integrated companies; small firms perform better than large companies, with higher labor productivity and lower unit labor costs; and state-owned enterprises (SOEs) have lower productivity (labor and capital) than private firms—and this gap is widening.

Second, Croatia lags behind its regional peers on entrepreneurship measures, suggesting lower economic dynamism. The entry density is below that predicted by its income, and its rates of high-growth firms and "gazelles"—a subset of high-growth enterprises up to five years old—are also below Europe and Central Asia (ECA) peers' rates.

Third, the lack of dynamism of the economy is confirmed in an analysis of firm entry and exit. In Croatia, only 5.5 percent of all firms were new to the market every year over 2008–12; in ECA peers the rate was 9–18 percent. Croatia also lags behind on exit: 6.5 percent versus 7–26 percent in ECA peers. When we look at net entry rates (entry minus exit), Croatia presented negative values, indicating that exit outpaced entry over 2008–12, reinforcing the view that Croatia is a stagnant economy with reduced creative destruction and inno-vativeness. There is some sector variation, however: although all macro sectors showed a negative entry effect, the process was less prominent in services. In addition, among only service activities there is evidence of strong firm dynamism for both knowledge-intensive (KI) services and other less knowledge-intensive (LKI) services.

Fourth, the contribution of firm dynamism (proxied by net entry) for productivity growth in Croatia is surprisingly negative. The economy saw a decrease of 2.88 percentage points in total factor productivity (TFP) over 2008–12. The decomposition of this variation in contributions from "survival," "start-up entry," "big-entry," and "exit" firms points to a negative contribution of the net entry effect, which is not what is expected: in principle, the net entry effect should be positive, reflecting the gains arising from the creation of new firms and the exit of obsolete ones. The negative contribution of the net entry effect suggests that the creative destruction process in Croatia is inefficient, as the market might be eliminating firms that are potentially productive (or conversely, preventing the entry of more efficient firms). The inefficiency of this process seems more pronounced among private firms, Continental region firms, and firms in the services sector (yet not all services subsectors had a negative performance, with KI service companies showing some untapped potential for productivity growth).

Croatia's poor export performance in the last few years (as seen in the previous chapter) stems partly from supply-side constraints and lack of competitiveness. Because the main factors in firm competitiveness—size, organization, technological capacity, and ability to operate in international markets—are ultimately related to firm productivity, this chapter explores stylized facts on productivity growth, drawing on firm-level data.

Exploring Firm Heterogeneity: Are There Regional, Export Status, Ownership, and Size Trade-Related Disparities?

Given their intrinsic characteristics, firms are self-evidently different from each other, which affects the aggregate pattern of productivity or any other output measure of a country. On aggregate measures of performance, even in expanding sectors, there are firms experiencing decline, while in expanding industries there are always outstanding expanding firms. Thus the aggregate pattern of performance—of a country or a sector—hides wide differences in performance among individual firms. Exploring firm heterogeneity is thus particularly important for policy makers wanting to prioritize investments or to explore complementarities among firm characteristics. To shed light on firm heterogeneity and explore the differential performance across several analytical criteria, we estimate the average difference of performance indicators across four categories—region, export status, size, and ownership (state-owned vs. private)—while controlling for (2-digit) sector differences (box 4.1).[2]

The results show that, except for unit labor costs, Adriatic region firms fare worse than firms in the Continental region (figure 4.1).[3] In 2008, labor productivity was 20.09 percent and capital productivity was 24.25 percent lower; however, this gap is slowly closing, especially in capital productivity (the gap narrowed by about 14 percentage points over 2008–12). Unit labor costs are consistently 8.65 percent higher in the Adriatic region.

Trade-integrated firms (i.e., firms that export and import) have better performance on all three indicators (figure 4.2). Also, firms that are either exposed to

---

Box 4.1 Sample Characteristics

The analysis was performed on a representative sample of 2,000 Croatian companies, obtained from the Croatian Financial Agency (FINA) database. For large companies, all companies with more than 250 employees in at least one of the analyzed years (2008, 2010, and 2012) were included in the dataset. The rest of the sample was divided roughly equally between small and medium enterprises; for each of these two subsets, a proportional stratified sample was constructed, with county and sector (2-digit) as control variables.

The sample comprised 729 small, 719 medium, and 552 large enterprises. In relation to the population of all Croatian firms, this means that the sample contains 100 percent of large firms, 55.5 percent of medium firms, and 0.97 percent of small firms. Within each of the three size groups, the sample is representative for sector and county.

Four main firm performance indicators were estimated: labor productivity, capital productivity, unit labor cost, and TFP.

$$Labor\ productivity = \frac{Real\ valued\ added\ (at\ factor\ cost), deflated\ by\ sector\ deflator}{(average)\ number\ of\ (full\ time)\ employees}$$

$$Capital\ productivity = \frac{Real\ valued\ added\ (at\ factor\ cost), deflated\ by\ sector\ deflator}{Real\ stock\ of\ capital\ (deflated\ by\ GDP\ deflator)}$$

$$Unit\ labor\ cost = \frac{Cost\ of\ personnel}{Real\ valued\ added\ (at\ factor\ cost), deflated\ by\ GDP\ deflator}$$

TFP is defined as the Solow residual of the production function and is estimated using the methodology of Levinsohn and Petrin (2003). A production function with output as the dependent variable and with capital, labor, and material costs as input variables was used. Specifically, output was proxied by real value added at factor cost; labor by (average) number of full-time employees; material inputs by real material costs; and capital by real stock of tangible fixed assets. To control for differences in production technologies across sectors, the TFP analysis estimates heterogeneous sector-specific (1-digit) production functions. Aggregation at this level (instead of the 2-digit level) is because some 2-digit sectors have too few observations for estimation using the above methodology.

---

international competition by exporting products or connected to global markets through direct imports (most likely of intermediate inputs) perform better than firms operating solely in the domestic market. Being neither an importing firm nor an exporting firm is thus the worst combination: in 2012, labor productivity of those firms was 51.76 percent lower and their capital productivity 27.42 percent lower, and their unit labor costs 27.72 percent higher, than among trade-integrated firms.

Small firms (10–49 employees) seem to perform better than large companies on all indicators (figure 4.3). Relative to large firms in 2012, they had 42.15 percent higher labor productivity, 51.47 percent higher capital

Figure 4.1 Percentage Difference of Firms in the Adriatic Region vs. Continental Region Average

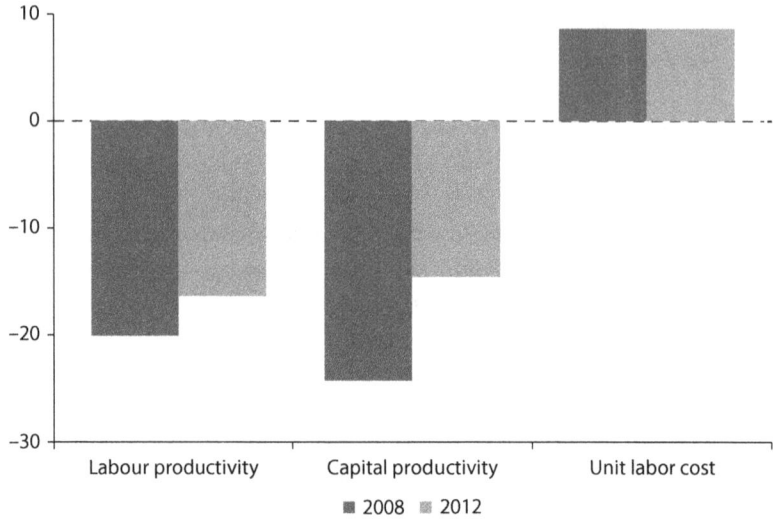

■ 2008 ■ 2012

Figure 4.2 Percentage Difference of Firms with Different International Exposure vs. "Exporters and Importers" Average

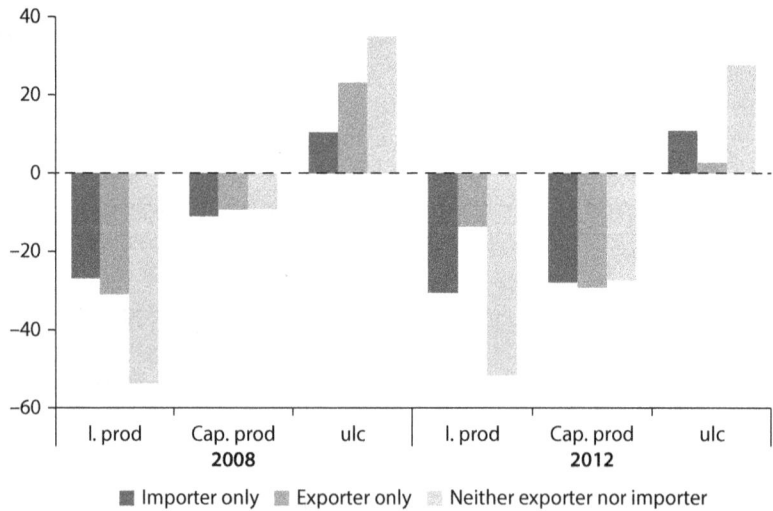

■ Importer only ■ Exporter only ■ Neither exporter nor importer

productivity, and 23.16 percent lower unit labor costs. Micro-enterprises (0–9 employees) seem to perform better on capital productivity, but fare worse on labor productivity and unit labor costs.

State-owned enterprises (SOEs) have lower productivity (labor and capital) and higher unit labor costs than do private firms (figure 4.4). In 2012, labor and capital productivity of private firms was 69.4 percent and 125.69 percent higher than of SOEs, respectively. This gap widened over the period.

Figure 4.3 Percentage Difference of Different-Size Firms vs. Large Firms (>250 Employees) Average

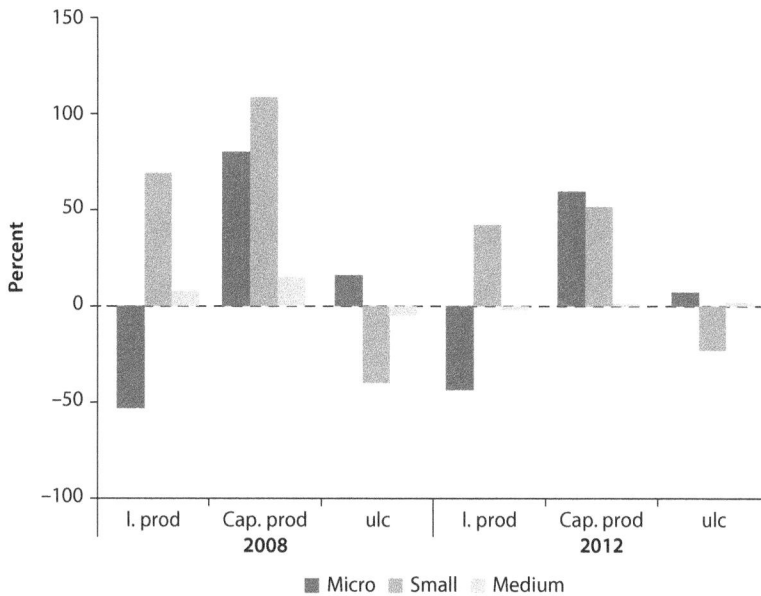

Figure 4.4 Percentage Difference of Private and Mixed Firms vs. SOE Average

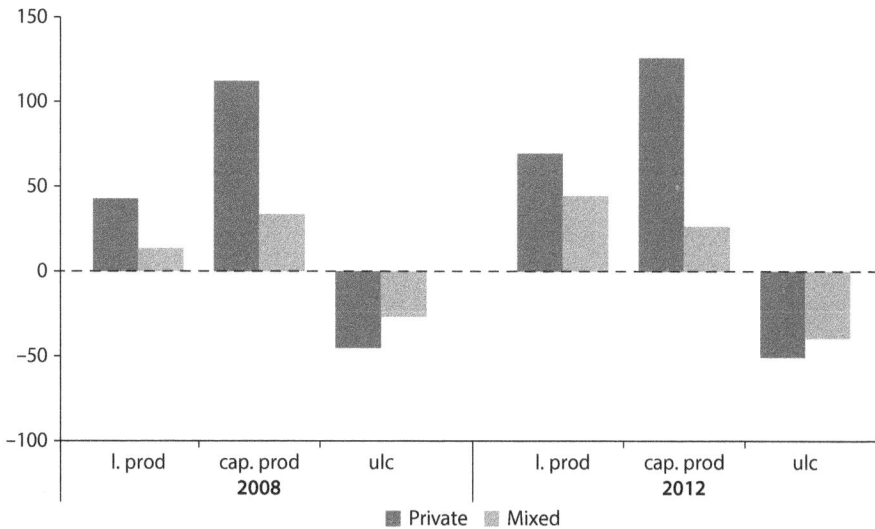

Entrepreneurship: High-Growth Firms and "Gazelles"

In firm dynamism and the intrinsic creative destruction process, an important feature is entrepreneurship—a fundamental driver of growth and development. Figure 4.5 displays the relationship between firm entry density (measured by the average annual number of new limited liability firms registered per 1,000

Figure 4.5 Entry Density and GDP per Capita, 2008–12

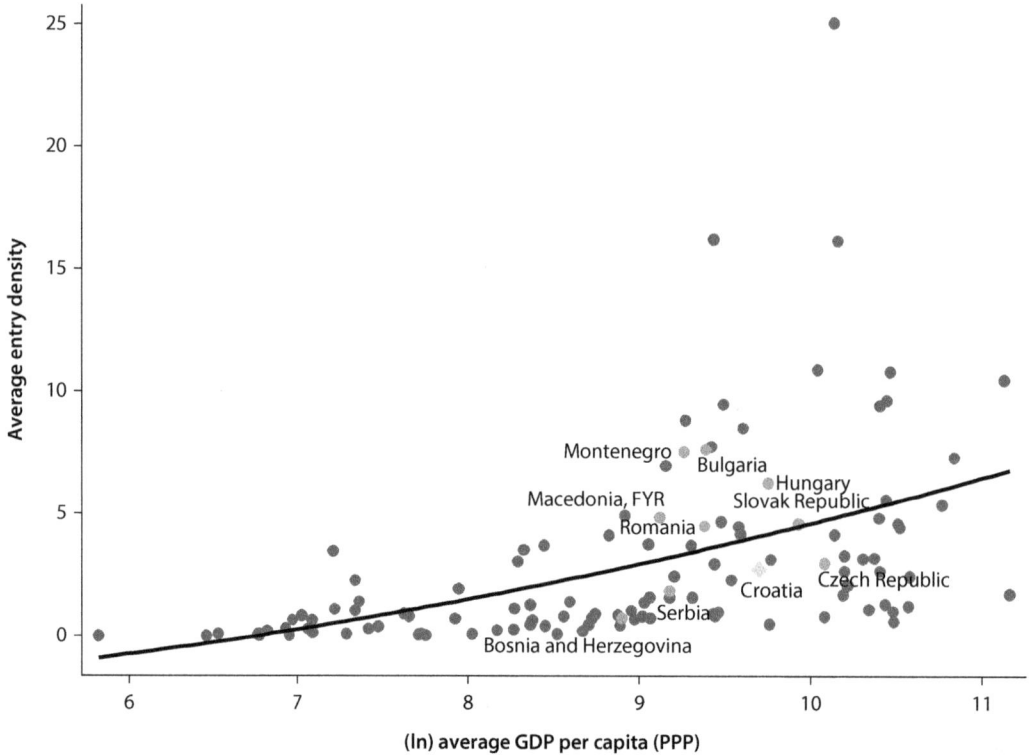

Source: Authors' elaboration based on data from World Development Indicators and World Bank Group Entrepreneurship Snapshots databases.

working-age people during 2008–12) and economic development (measured by average per capita income for the same period) across 117 countries. As expected, entry is positively associated with GDP per capita, but Croatia's rate of entry density is far below what would be predicted by its income. This result contrasts with some Europe and Central Asia (ECA) peers, such as Bulgaria, Hungary, the former Yugoslav Republic of Macedonia, and Romania, which are above the predicted line.

Also important is the entry of high-growth firms and the subset of "gazelles." Based on the Organization for Economic Co-operation and Development (OECD) definition, high-growth enterprises are those with average annualized growth in employees (or turnover) greater than 20 percent a year, over a three-year period, and with 10 or more employees at the beginning of the observation period. ("Gazelles" are a subset of high-growth enterprises up to five years old.)

These types of firms are of particular importance for two reasons. First, by their extraordinary growth, they can make the largest contribution to net job creation, despite typically representing a small proportion of the business population (Henrekson and Johansson 2010, for example). Second, because their success often comes from innovation—such as a product or process;

Figure 4.6 High-Growth Firms' Rate, 2010, Measured by Turnover Growth

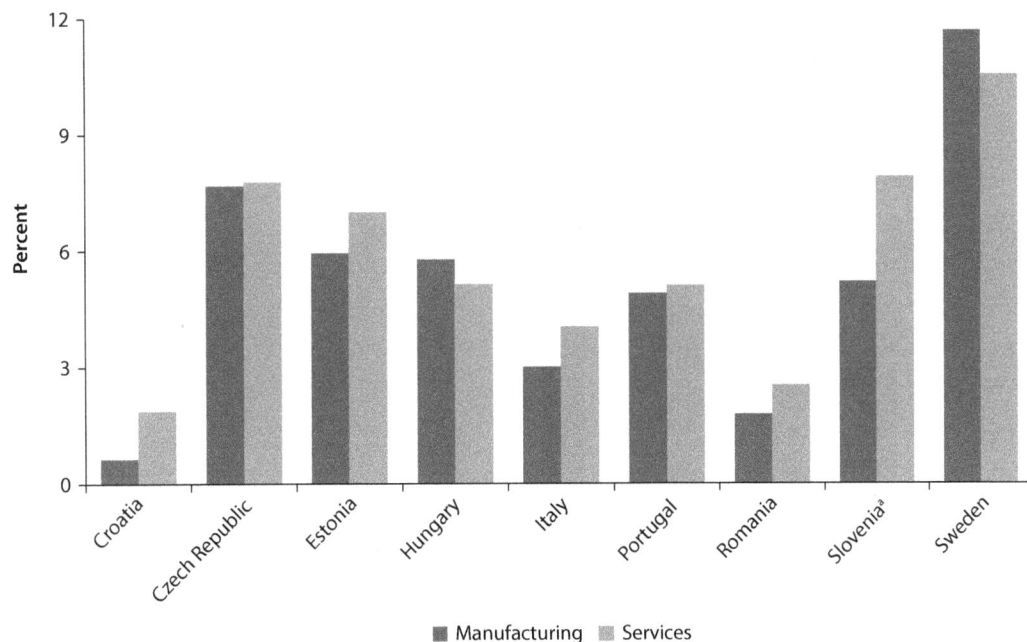

Source: Authors own' elaboration based on Entrepreneurship at a Glance 2013, OECD and FINA data (for Croatia).
Note: Rate is defined as the number of high-growth enterprises as a percentage of the population of enterprises with 10 or more employees.
a. For Slovenia, data are for 2009.

or innovative approaches to marketing, distribution, or organization; or from entering a new market—these types of firms (particularly gazelles) outperform average industry growth in a Schumpeterian sense[4]: they combine existing factors of production in a new way and thus produce an innovation that enables them to outperform the market.

Yet Croatia seems to lag behind selected European countries on high-growth firms and gazelles (figures 4.6 and 4.7)—although these findings have to be treated with caution because the numbers were not drawn from population figures.

Firm Dynamics and Productivity Growth: Firm Entry and Exit

The objective of this section is to explore the links between productivity growth [proxied by total factor productivity (TFP) growth] and firm dynamics. At the aggregate (and industry) level, productivity growth results from the combination of three main components: the "within component" accounts for the productivity gains within existing firms; the "reallocation component" measures the productivity growth coming from the reallocation of labor and capital (and inputs, more generally) across existing firms; and the "net entry component" reflects the

Figure 4.7 Gazelles' Rate, 2010, Measured by Turnover Growth
Percent

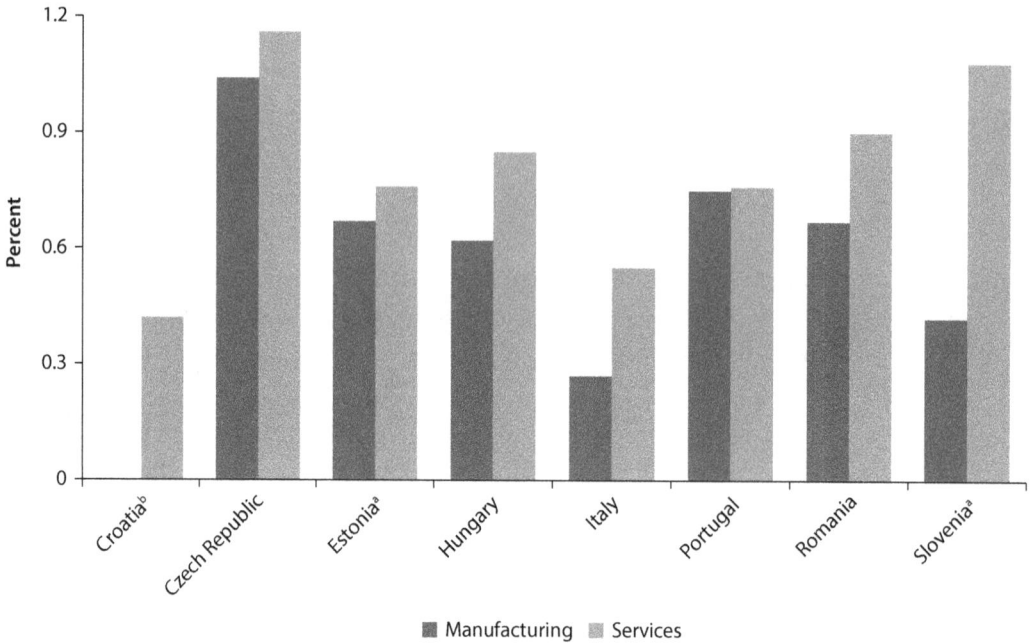

Source: Authors own' elaboration based on Entrepreneurship at a Glance 2013, OECD and FINA data (for Croatia).
Note: Rate is defined as the number of gazelles as a percentage of the population of enterprises with 10 or more employees.
a. For Estonia and Slovenia, data are for 2009.
b. For Croatia, we did not identify any gazelles in 2010 (a sample of firms was used).

productivity growth resulting from the entry of new, more productive firms and the exit of obsolete ones.

While productivity growth is largely driven by the continuous process of restructuring and upgrading by existing firms,[5] the entry of firms and the exit of obsolete units—captured by the net entry (or firm-churning) effect—is also important (box 4.2). This component is key for business dynamism and economic growth because it drives the creative destruction process that facilitates innovation or adoption of new technology, helping shift resources from less productive firms to more productive firms.

For our analysis, we introduce a two-step approach, in light of the methodology used by Geishecker, Gorg, and Taglioni (2009). We first identify groups of entering, surviving, and exiting firms; then we calculate the contribution of each of those groups to aggregate domestic productivity change. On the basis of our final sample from FINA data, we identify four groups of firms: surviving firms (S), new start-ups (NSU), new big entries (NBE), and exit firms (X). Surviving firms were active in both 2008 and 2012, whereas exit enterprises[6] were active only in 2008.[7] NSU are firms incorporated after 2008 but sampled only in 2012. NBE are firms that were incorporated before 2008 and that were active in both

Box 4.2 Croatia Shows Low Rates of Firm Entry and Exit

We estimated the aggregate total factor productivity (TFP) change for 2008–12 using Equation 1, where $\pi_{i,t}$ denotes firm i's productivity (natural logarithm of TFP or labor productivity) in year t, $\theta_{ij,t}$ is the share of firm i in total employment (turnover) of a NACE Rev. 2 code 1-digit sector j in a given year t, and $\Delta\Pi_t$ denotes aggregate productivity (TFP or labor productivity) change over 2008–12. Variables on the right hand side of Equation 1 stand for the contribution of each group of firms to the overall productivity change: the first term denotes the overall growth contribution of surviving firms, the second term denotes the growth contribution of new market entries as start-ups; similarly, the third term denotes the contribution of market entries as new bigs; while the last term represents the growth contribution of market exits.

$$\Delta\Pi t = \sum_{i\in S}\left(\theta_{ij,2012}\times\pi_{i,2012}-\theta_{ij,2008}\times\pi_{i,2008}\right)+\sum_{i\in NSU}\left(\theta_{ij,2012}\times\pi_{i,2012}\right)$$
$$+\sum_{i\in NBE}\left(\theta_{ij,2012}\times\pi_{i,2012}-\theta_{ij,2008}\times\pi_{i,2008}\right)-\sum_{i\in X}\left(\theta_{ij,2008}\times\pi_{i,2008}\right)\tag{1}$$

For values of $t \in$ {2008, 2012}, $\theta_{i,j,t}$ for employment weighting is calculated as follows:

$$\theta_{i,j,t}=\frac{l_{i,t}}{\sum_{i\in j}l_{i,t}}$$

where $l_{i,t}$ is the employment of firm i in time t. In addition to weighting by employment, in Equation 2 we can also weight by turnover. In that case, instead of $\theta_{ij,t}$ we use $\mu_{ij,t}$, $t \in$ {2008, 2012}, which is defined in the following way:

$$\mu_{i,j,t}=\frac{\tau_{i,t}}{\sum_{i\in j}\tau_{i,t}}$$

where $\tau_{i,t}$ is the turnover of firm i in time t.[a]

 This basic decomposition is then extended by adding more criteria to distinguish firms by region, size, ownership, macro sector, and international exposure.

 On the right-side components, the net entry contribution, defined as the difference between entry and exit cohorts, is especially important. As highlighted by Bartelsman et al.

Table B4.2.1 Equation (1) Decomposition, 2008–12
Percentage points

	Productivity metric	TFP[a]		Labor productivity[a]	
	Weighted by	Employment	Turnover	Employment	Turnover
Productivity change	$\Delta\Pi_{t,t-k}$	−2.88	−6.56	−3.67	−7.22
Survivals	$\sum_{i\in S}(\theta_{ij,t}\times\pi_{i,t}-\theta_{ij,t-k}\times\pi_{i,t-k})$	2.87	0.96	2.99	0.27
New big entry	$\sum_{i\in NBE}(\theta_{ij,t}\times\pi_{i,t}-\theta_{ij,t-k}\times\pi_{i,t-k})$	0.60	0.33	0.53	0.29
New start-ups	$\sum_{i\in NSU}(\theta_{ij,t}\times\pi_{i,t})^{**}$	1.25	1.05	1.11	0.99
Exit	$\sum_{i\in X}(\theta_{ij,t-k}\times\pi_{ij,t-k})^{**}$	−7.60	−8.91	−8.32	−7.22

a. Both TFP and labor productivity were used in natural logarithm form. These are given in (ln) productivity units; their difference is % change.

box continues next page

Box 4.2 Croatia Shows Low Rates of Firm Entry and Exit *(continued)*

(2009), the within-country variations of the net entry effect is particularly important for policy purposes because they may reflect poor market structure and institutions that distort the contribution of creative destruction.[b]

The box table shows the productivity decomposition according to two productivity measures and two different weighting options to compute Equation 1. Results are quite robust. Productivity declined for both labor productivity and TFP (regardless of the weighting alternative); further, in all cases the largest positive contribution comes from survivals—consistent with the literature—while the net-entry effect is shown to be negative with the exit contribution largely outpacing the (total) entry effect (new entry and new big entry). Further analysis and results in this chapter are for the TFP case weighted by employment.

a. Equation 1 was calculated only for firms for which we had observations for both 2008 and 2012 (or just 2008 or 2012 if firms are classified as exits or new start-ups, respectively), thus further reducing the sample to 1,195 firms.
b. On the other hand, a cross-country comparison of the contribution of this effect is more challenging due to potential measurement and interpretation problems (associated with firm turnover) that cloud rank orderings across countries.

Table 4.1 Entry and Exit Rates, Croatia, 2008–12, and ECA Peers, 2008–11

Country	Entry	Exit	Net entry
Bulgaria	14.54	9.62	4.92
Czech Republic	9.18	8.96	0.22
Latvia	16.57	13.27	3.30
Lithuania	18.61	26.82	8.21
Hungary	9.83	10.92	1.09
Poland	13.07	10.42	2.65
Romania	10.92	14.19	3.27
Slovak Republic	14.86	11.56	3.30
Slovenia	10.89	7.53	3.36
Croatia	5.50[a] ("start-ups")	6.50	1.00

Source: World Bank staff elaboration based on FINA data (for Croatia) and Eurostat data.
Note: For all countries except Croatia, values are averages of Eurostat data for 2008–11 (not 2008–12). Eurostat's yearly measure of entry is defined as number of enterprise births in the reference period (t) divided by the number of enterprises active in t. The yearly exit rate is defined in Eurostat as number of enterprise deaths in the reference period (t) divided by the number of enterprises active in t. For both entry and exit rates from Eurostat, the sector coverage is "Business economy except activities of holding companies."
a. The rate for "big entry" is 2.5.

2008 and 2012, and sometime in that period they grew until they had at least 50 employees.

Healthy market economies exhibit high rates of firm entry and exit, whereas the economy of Croatia does not. This churn indicates an intense process of creative destruction, in which a significant number of businesses start or close operations (Foster, Haltiwanger, and Krizan 2001; Bartelsman et al. 2009). Table 4.1 suggests that Croatia had little dynamism by that criterion over 2008–12.

In the peer countries, 9–18 percent of all firms are new to the market every year, whereas in Croatia only 5.5 percent are new. (Croatia's figure considers only the real entry firms, i.e., the start-ups; the rate is 8 percent if we "super estimate" the entry process by also including firms that have transitioned from micro/small companies to medium companies over 2008–12.) Croatia's lagging performance is even more pronounced on the exit process: 6.5 percent versus 7–26 percent in peer countries. On net entry rates (entry minus exit), two results emerge. First, a common pattern across countries is that net entry is far less important than the gross entry and exit of firms, which means that changes in total number of active firms are generally small despite gross entry and exit. This confirms previous findings (Bartelsman et al. 2009, for instance) and suggests that the entry of new firms in the market is part of a process of search and experimentation (rather than a response to sub- or supra-normal profits).[8] The second result is that while in some countries, such as Bulgaria, Latvia, Poland, the Slovak Republic, and Slovenia, firm entry largely outpaces firm exit (essentially related to the transition to market economies), in other countries, Croatia included, the exit rate is higher than the entry rate. This reinforces the perception that Croatia is a stagnant economy with a minimized creative destruction process and reduced innovativeness, which in turn undermines export diversification.

In Croatia, firm dynamics vary more across sectors than across regions. Net entry effects are negative for the Continental and Adriatic regions and quite similar in size (table 4.2). Variation within sectors, however, is higher: table 4.3 shows that apart from mining and quarrying—where exit exactly cancels entry—the sector presenting the largest flow of firms entering is services, where, on average, 7.58 percent of firms are new to the market every year. Except for construction,

Table 4.2 Exit and Entry Rates in Croatia by Region, 2008–12
Percent

Region	Start-up entry rate	New big entry rate	Exit rate	Net entry (start-up entry—exit)
Continental	5.01	2.76	6.24	1.2
Adriatic	6.64	1.9	7.11	0.5
Total (Croatia)	**5.5**	**2.5**	**6.5**	1.0

Note: Number of firms, N = 1,400.

Table 4.3 Exit and Entry Rates in Croatia by Macro Sector, 2008–12
Percent

Macro sector	Start-up entry rate	New big entry rate	Exit rate	Net entry (start-up entry—exit)
Agriculture, forestry, and fishery products	2.63	7.89	5.26	2.6
Mining and quarrying	7.69	0.0	7.69	0.0
Manufacturing	2.67	2.67	4.8	2.1
Electricity, water supply	0.0	1.47	1.47	1.5
Construction	4.69	2.34	2.34	2.4
Services	7.58	2.31	8.48	0.9
Total (Croatia)	**5.5**	**2.5**	**6.5**	**1.0**

Note: Number of firms, N = 1,400.

Table 4.4 Exit and Entry Rates in Croatia in Services Macro Sectors Only, 2008–12
Percent

Services macro sector	Start-up entry rate	New big entry rate	Exit rate	Net entry (start-up entry—exit)
KI market	10.17	0.00	6.78	3.39
KI high-tech	20.00	4.00	8.00	12.00
KI financial	0.00	0.00	25.00	25.00
KI other	2.27	2.27	6.82	4.55
LKI market	6.38	2.81	8.63	2.25
LKI other	15.38	0.00	0.00	15.38
Total (services only)	**7.58**	**2.31**	**8.48**	**0.90**

Note: Number of firms, N = 778. Examples of services macro sectors are given in the discussion on figure 4.13. KI = knowledge-intensive; LKI = less knowledge-intensive.

all sectors show a negative entry effect (start-up entry minus exit), but again, this process seems less prominent in services.

However, some exceptions within services stand out. Among six main subcategories within services—as defined by Eurostat—according to knowledge intensity (table 4.4), the churning rate for KI high-tech services (telecoms, computer-related activities, R&D activities, and so on) is highly positive, which is expected given these are notably young activities with strong innovative drivers. Other less KI services also present a high and positive churning rate, although the low knowledge content of this type of activity (proxied by personal services, for instance) points to a worrisome trend.

Based on the empirical exercise in box 4.2, figure 4.8 displays the decomposition of aggregate TFP change in 2008–12. The greatest positive contribution comes from surviving firms (2.87), followed by start-ups (1.25), and new big entry firms (0.6). These three components together, however, were not enough to offset the loss in TFP due to exiting firms of more than 7.59 percentage points, amounting to a total fall in TFP of 2.88 percentage points.

Performing the same decomposition exercise but from a regional perspective shows that more than half (56 percent) of the fall in TFP comes from the Adriatic region (figure 4.9). The results also show that the largest negative contribution for productivity growth comes from the net entry effect of Continental region firms, which suggests a more inefficient market selection mechanism there.

Figure 4.10 shows the TFP change decomposition by ownership of enterprises. Surviving private sector firms present the largest positive contribution. The net entry effect is, though, negative and larger among private firms, suggesting that market selection mechanisms are particularly inefficient for these types of firms. And unsurprisingly, survivor SOEs contribute negatively to productivity growth.

Figure 4.11 shows TFP change decomposition by export status. Four categories are considered according to export status in 2008 and 2012: firms that exported in both periods; firms that did not export in either year; firms that moved from nonexporter status to exporter status; and firms that moved from

Figure 4.8 TFP Change Decomposition by Type, 2008–12

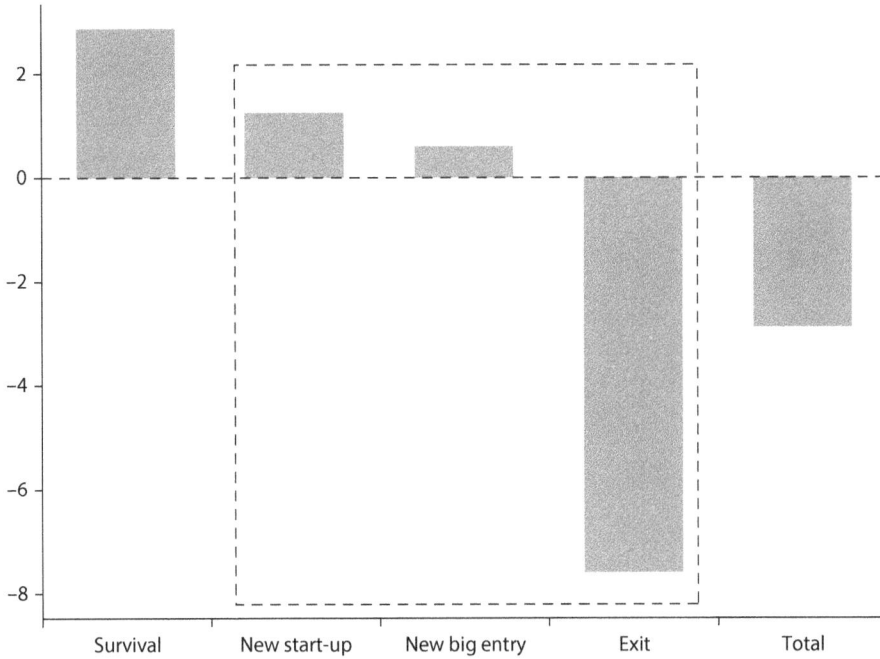

Note: TFP is used in natural logarithm form.

Figure 4.9 TFP Change Decomposition by Region, 2008–12

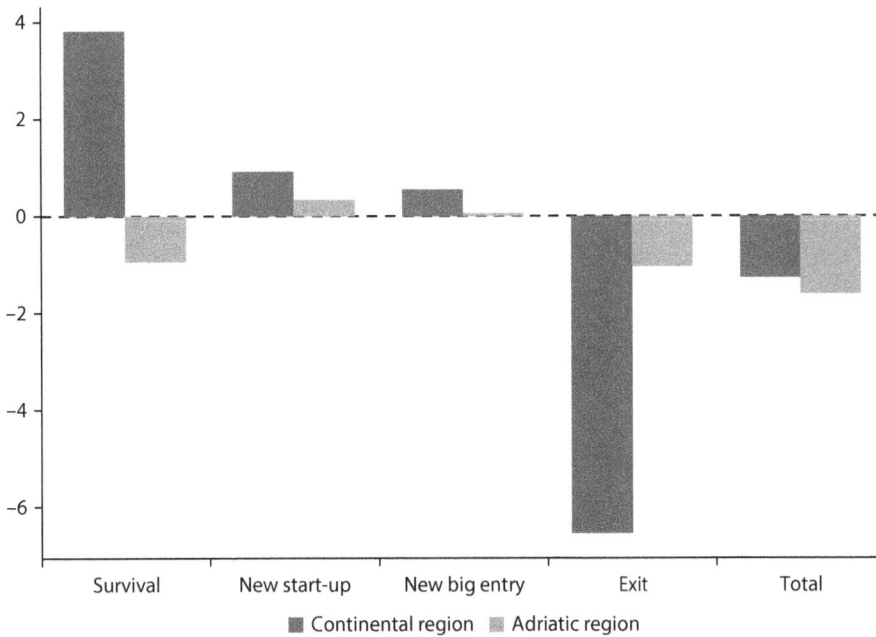

■ Continental region ■ Adriatic region

Note: TFP is used in natural logarithm form.

Figure 4.10 TFP Change Decomposition by Ownership, 2008–12

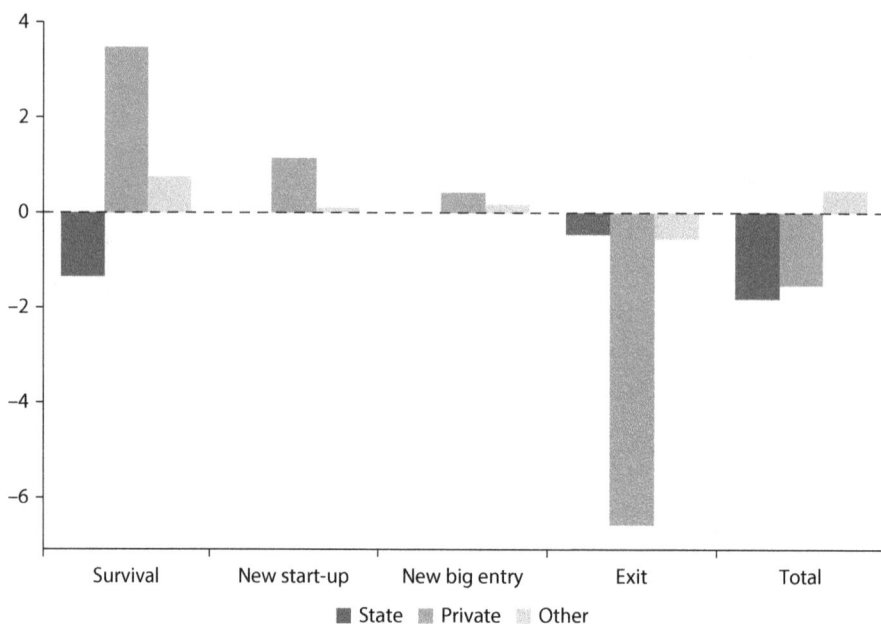

Note: TFP is used in natural logarithm form.

Figure 4.11 TFP Change Decomposition by Type and by Export Status, 2008–12

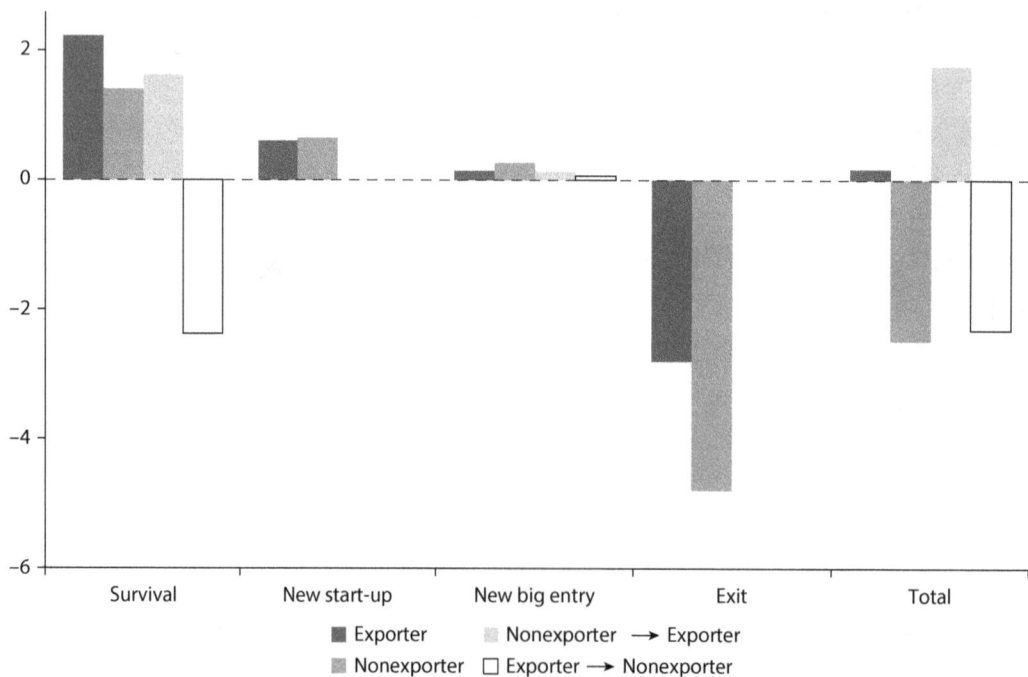

Note: TFP is used in natural logarithm form.

exporter status to nonexporter status. Results show that productivity growth depends essentially on export activity (with no surprise there), because both exporters and firms that transited from nonexporter status to exporter status contributed positively to overall TFP performance.

Figure 4.12 displays the results of TFP growth decomposition by sector. The negative performance of services activities is the main sector driver behind the falling performance of TFP in Croatia in 2008–12; overall shrinkage in services productivity was 3.93 percentage points, or worse than the contraction in productivity of the whole economy (down 2.88 percentage points). On both survival and net entry contributions, services emerge again as the main driver: services firms present the highest positive contribution to productivity performance among surviving firms; and, though the entry of new services firms is generally positive, it was not large enough to outweigh the negative effect from the exit of services firms. That the net entry effect is particularly negative for services firms suggests that inefficient market selection is especially pronounced for such firms.

Still, not all services firms had negative performance. Figure 4.13 presents the TFP growth decomposition across six main subcategories within services based on knowledge intensity: knowledge-intensive (KI) market services; KI high-tech services; KI financial services; KI other services; less KI market services; and less

Figure 4.12 TFP Change Decomposition by Type and by Macro Sector, 2008–12

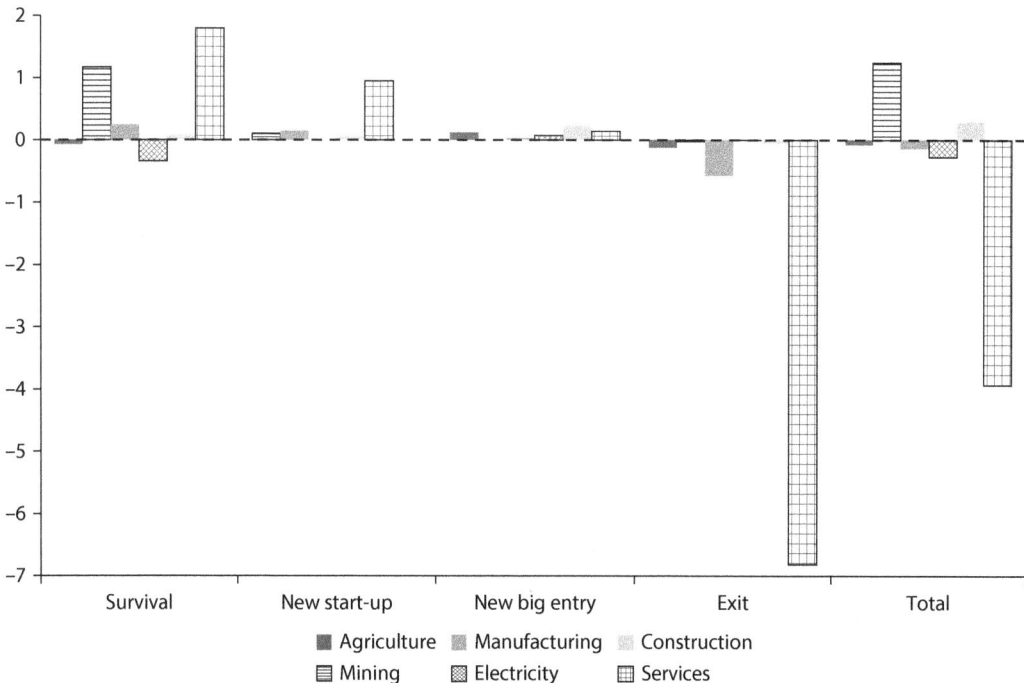

Note: TFP is used in natural logarithm form.

Figure 4.13 TFP Change Decomposition by Type and by Services Macro Sector Group, 2008–12

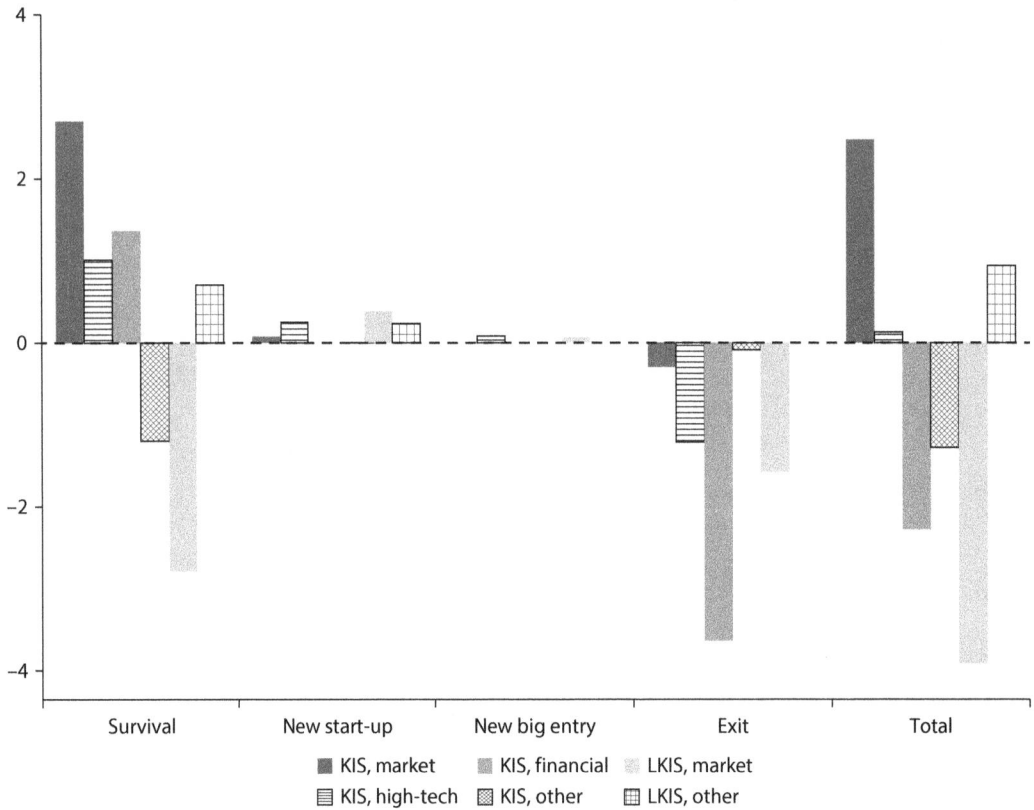

Note: TFP is used in natural logarithm form.

KI other services. Some groups of services have outperformed: KI market services (such as water, air, real estate), KI high-tech services (telecoms, computer-related activities, R&D), and other less KI services (such as personal services) contributed positively to overall services productivity performance. The figure also reveals some untapped potential from entry cohorts, but the productivity contribution from start-ups and dynamic micro/small firms (NBE) is particularly low among KI services.

Factors That Can Affect Productivity

Firm dynamism (proxied by net entry) had a negative contribution to productivity growth over 2008–12. This implies that the creative destruction process has not been working properly, as the market might be eliminating firms that are potentially productive (or conversely, preventing the entry of more efficient firms). What are the likely reasons? The answer can be found in the existing anticompetitive regulation and policies in the period, notably the state aid system; the specific product market policies for services; and the barriers to firm exit and expansion, proxied, respectively, by insolvency resolution and contract enforcement.

State aid is, sometimes, necessary for a well-functioning and equitable economy as well as for the public interest. However, it can also create distortive effects in the market by impairing competitive neutrality. Evidence suggests that Croatia still had pervasive state aid in 2011 (the last year with data). For instance, total state aid (excluding that to railways) in that year came to 2.47 percent of GDP,[9] or far higher than the EU-27 average of 0.51 percent of GDP (figure 4.14).[10] In specific sectors the aid was much higher than the EU-27 average: 1.08 percent of GDP against 0.07 percent (figure 4.15). Three main sectors accounted for the largest part of sector aid: transport (0.42 percent of GDP), public service broadcasting (0.35 percent), and shipbuilding (0.21 percent). The most common aid instruments were, in order, grants, guarantees, participation in equity capital, and soft loans. Public service compensation—given to firms entrusted with operating services of general economic interest—is another instrument. This is pronounced for public service broadcasting (essentially

Figure 4.14 Total Noncrisis State Aid

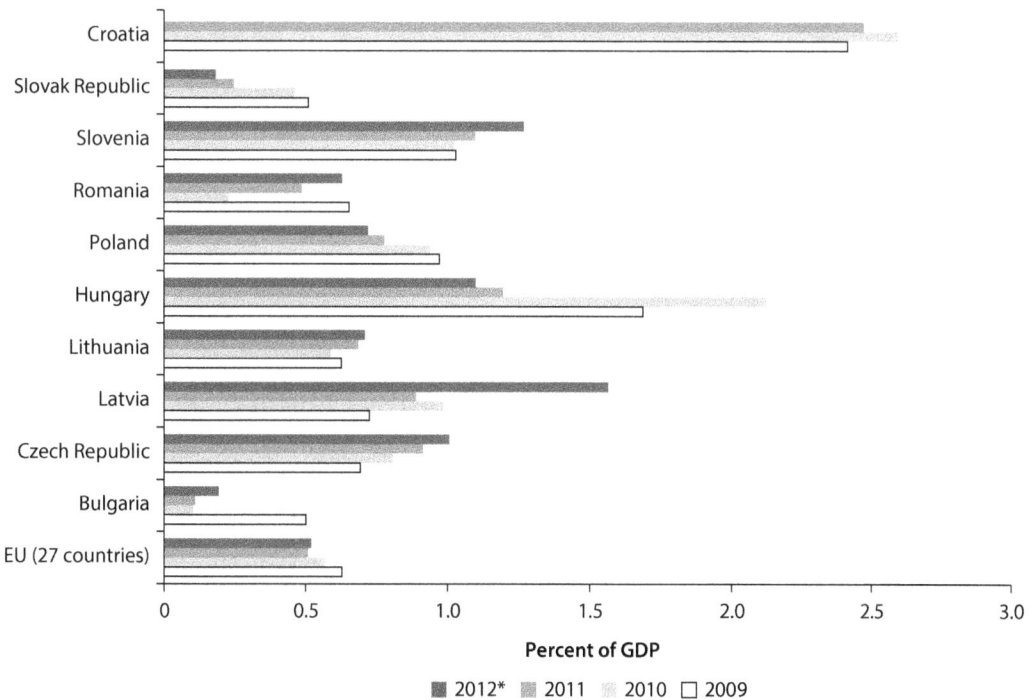

Source: World Bank staff elaboration based on data from European Commission, DG Competition (European Commission 2012. Facts and Figures on State Aid in the EU Member States—2012 Update) and Croatia Competition Agency (Croatian Competition Agency. 2012. *Annual Report on State Aid for 2011*).
Note: No data were available for Croatia in 2012.

Figure 4.15 Noncrisis State Aid: Selected Aid Categories for Croatia, 2011, and EU-27, 2012

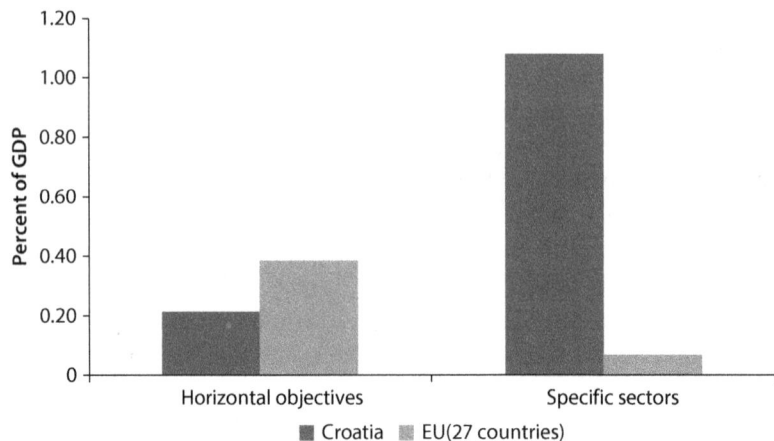

Source: World Bank staff elaboration based on data from European Commission, DG Competition (European Commission 2012. Facts and Figures on State Aid in the EU Member States—2012 Update) and Croatia Competition Agency (Croatian Competition Agency 2012. Annual Report on State Aid for 2011).
Note: No data was available for Croatia in 2012.

Hrvatska Radiotelevizija, the Croatian Radio–TV company) and, to a lesser degree, for transport (land maritime and air transport) (figure 4.16).

As part of the EU accession process and as required by EU legislation, Croatia has aligned its rules on state aid (as well as those on antitrust and mergers) with the *acquis communautaire*. For this reason, some of the state aid practices reflected in 2011's data might have changed. As the state aid metrics have not yet been updated, it is unclear whether the amount of aid granted by the government has in fact contracted and converged to EU levels. In any case, recent perception data from the World Economic Forum suggest that competition in the domestic market is still weak, with Croatia faring poorly against ECA peers (figure 4.17).

Besides state aid, factors related to product market policies can help explain why competition in Croatia was largely hobbled in 2008–12, possibly preventing the most efficient players from emerging (and the less efficient ones from exiting). Using the OECD indicators for Product Market Regulation and associated methodology, World Bank (2009) showed that Croatia was, in 2008, the sixth most restrictive economy among 30 countries on conditions for product market competition (and more restrictive than predicted by its development level). Though much has changed since 2008, following the required adoption of norms from the EU's *acquis communautaire*, some misalignments may still exist, especially in services. These can be particularly problematic because competition in some nonmanufacturing

Figure 4.16 Sector-Specific Aid, 2009–11

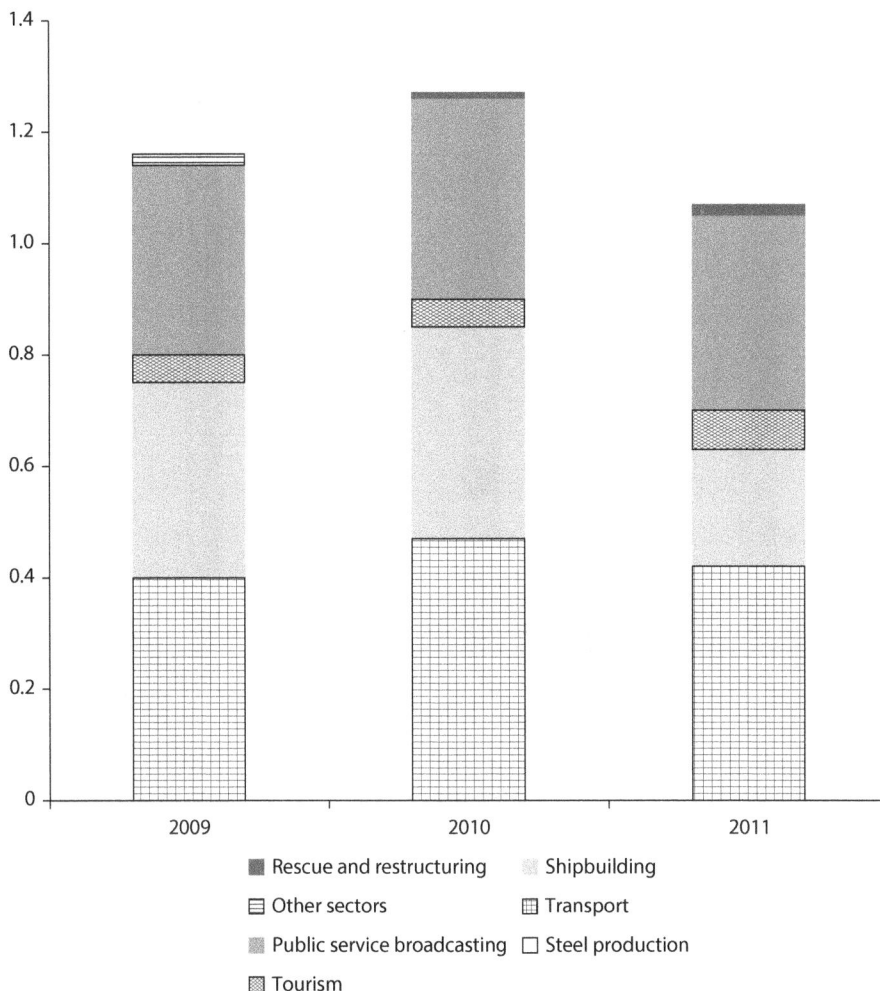

Source: World Bank staff elaboration based on data from Croatia Competition Agency (*Annual Report on State Aid for 2011*).
Note: No data available for Croatia in 2012.

sectors—energy (gas and electricity), transport (airlines, railways, road freight), communications (post, telecoms), and professional services—is especially relevant for the efficiency of the whole economy. Shortcomings in product market regulation for services hit the cost structure of firms that use the output of services as intermediate inputs in production. State control over services, particularly through involvement in business operations, may still be significant. Although privatization has shown some recent successes, as proxied by the recent restructuring of the shipbuilding sector in 2013, in preparation for EU accession (European Commission 2013; IMF 2014),

Figure 4.17 Intensity of Local Competition, 2013–14

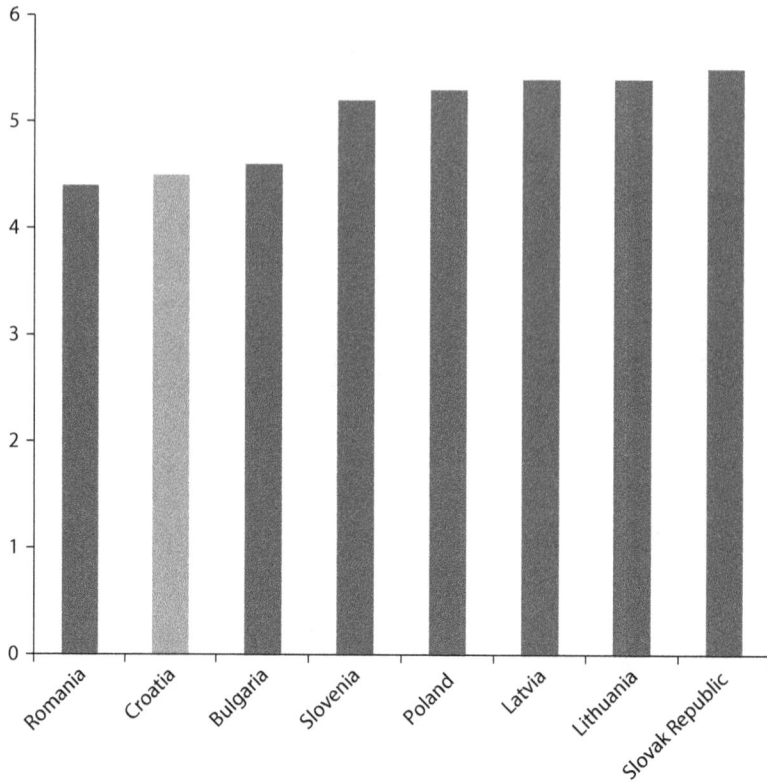

Source: World Bank staff elaboration based on *Global Competitiveness Report 2013–2014* (World Economic Forum 2013).

the use of coercive regulation—as opposed to incentive-based regulation—is still present economywide and in specific services subsectors.

A weak business environment is one of the reasons behind the overall poor firm exit and expansion performance in Croatia. Although the country has been an active reformer in some administrative areas over the last decade, the country still under performs on several measures related to the ease of doing business. Over the past decade, the country has managed to improve the regulatory environment in a number of key areas, such as starting a business or paying taxes. Nevertheless, despite these improvements, Croatia still ranks low (65th place) for the overall ease of Doing Business[11]—the second poorest performance by an EU member. According to the most recent DB data (2015), the areas which drive the disappointing performance include a number of key regulatory issues, such as dealing with construction permits, resolving insolvency, or getting credit (see figure 4.18). Meanwhile, most EU peers, including Estonia, Latvia, Lithuania, and Slovenia, perform better than Croatia across the board (i.e., on almost all Doing Business indicators).

Figure 4.18 Measures of the Ease of Doing Business (Distance to Frontier)

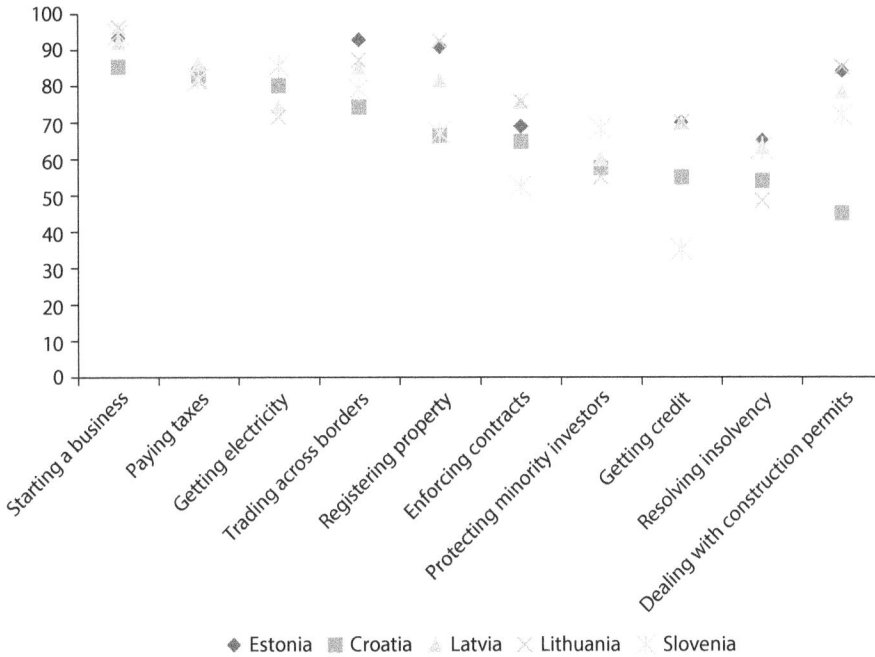

◆ Estonia　▦ Croatia　▲ Latvia　✕ Lithuania　✕ Slovenia

Source: Elaboration based on Doing Business 2015 indicators.
Note: The distance to frontier measure shows how far an economy is from the best performer (100) on each Doing Business indicator.

Notes

1. Inputs for this chapter came from the World Bank Policy Research Working Paper, "Stylized Facts on Productivity Growth: Evidence from Firm Level Data in Croatia," by Mariana Iootty, Paulo Correa, Sonja Radas, and Bruno Škrinjarić (2014).

2. For each analytical category (region, export status, size, and ownership), we estimate an OLS model of log performance indicators on the respective criteria and sector dummies (NACE Rev. 2 code 2-digit sectors).

   The estimated equation was of the following form:

   $$y_t = \alpha_t + \beta_t X_t + \sum_{\substack{j=1 \\ j \neq 28}}^{J} \gamma_{jt} nace\_2d_{jt}$$

   where y denotes natural logarithm of performance indicators used (labor productivity, capital productivity, and unit labor cost), X_t is the vector of dummies used for different categories (region/export status/size/ownership), and *nace_2d* is a vector of dummy variables for each NACE Rev. 2 code 2-digit sectors (except sector 28, the reference sector). Figures 4.1–4.4 show the results.

3. According to the Nomenclature of Territorial Units for Statistics (NUTS) of the EU, the regions at NUTS 2 level (nonadministrative) are the Continental and Adriatic regions. The counties in these regions are listed in annex 1.

4. The Schumpeterian process of "creative destruction" is crucial to fostering innovation and, therefore, economic growth. This process features the entry of new innovative firms in the market, and the exit of uncompetitive, incumbent firms. This entry-and-exit dynamic reallocates factors of production across firms, from unproductive companies to productive firms, increasing aggregate productivity. Entry into innovation activities can occur in two ways: vertical entry or horizontal entry. Vertical innovations are related to the upgrading of the quality of existing products and services; horizontal inventions are associated with the creation of new goods and services. Entry always entails a fixed cost. However, vertical innovations tend to cost less, as they relate to incremental innovations. But horizontal inventions are usually expensive, as they typically involve a breakthrough.

5. The empirical literature presents vast evidence pointing to the within component as the main driver of productivity growth (see Bartelsman et al. 2009).

6. Firms might, however, still exist without performing business activities.

7. The dataset has 160 companies for which 2012's financial data are missing, which would classify them as an "exiter." However, to make sure this is not caused by incompleteness in the FINA data, we checked all these companies manually using other sources of information, like Poslovna Hrvatska and web pages of various business associations. We discovered 51 companies still active in 2012, and therefore moved them from exiters to survivors.

8. See Audretsch (1995) for a theoretical explanation. In sum, if firm turnover (entry and exit) was driven by profit made in the market, this should have a sizable effect on the total number of firms active in the market.

9. Croatian Competition Agency. 2012. *Annual Report on State Aid for 2011*. Zagreb, Croatia: Croatian Competition Agency. http://www.aztn.hr/uploads/documents/tn/godisnja_izvjesca/godisnje_izvjesce_DP_2011.pdf.

10. Croatian Competition Agency. 2012. *Annual Report on State Aid for 2011*. Zagreb, Croatia: Croatian Competition Agency. EU-27 average is from European Commission (DG Competition European Commission 2012. *Facts and Figures on State Aid in the EU Member States—2012 Update*. Commission Staff Working Document SWD (2012) 443 final. Brussels: European Commission.)

11. Out of 189 economies.

Bibliography

AMECO database (accessed November 5, 2013), http://ec.europa.eu/economy_finance/db_indicators/ameco/zipped_en.htm.

Audretsch, D. B. 1995. *Innovation and Industry Evolution*. Cambridge, MA: MIT Press.

Bartelsman, E., J. Haltiwanger, and S. Scarpetta. 2009. "Measuring and Analyzing Cross-Country Differences in Firm Dynamics." In *Producer Dynamics: New Evidence from Micro Data*, edited by T. Dunne, J. B. Jensen, and M. Roberts. National Bureau of Economic Research. Chicago: University of Chicago Press.

Biznet database (accessed January 25, 2013) http://www1.biznet.hr/HgkWeb/do/extlogon.

Business Croatia [Hrv. Poslovna Hrvatska] database (accessed January 13, 2014), http://www.poslovna.hr/.

Croatian Bureau of Statistics (accessed January 13, 2014), http://www.dzs.hr/default_e .htm.

Croatian Competition Agency. 2012. *Annual Report for 2011*. Zagreb: Republic of Croatia.

Croatian Financial Agency (FINA) database. http://www.fina.hr/Default.aspx?sec=1134.

European Commission. 2012. *Facts and Figures on State Aid in the EU Member States—2012 Update*. Commission Staff Working Document SWD (2012) 443 final. Brussels: European Commission.

———. 2013. "Non-crisis State Aid, Excluding Railways—Million EUR, % of GDP." In *State Aid Scoreboard*. http://ec.europa.eu/eurostat/tgm_comp/table.do?tab=table&init =1&language=en&pcode=comp_ncr_xrl_01&plugin=1.

Eurostat database (accessed January 13, 2014), http://epp.eurostat.ec.europa.eu/portal /page/portal/statistics/search_database.

Foster, L., J. Haltiwanger, and C. J. Krizan. 2001. "Aggregate Productivity Growth: Lessons from Microeconomic Evidence." In *New Developments in Productivity Analysis*. National Bureau of Economic Research. Chicago: University of Chicago Press.

Geishecker, I., G. Gorg, and D. Taglioni. 2009. "Characterizing Euro Area Multinationals." *The World Economy* 32 (1): 49–76.

Henrekson, M., and D. Johansson. 2010. "Gazelles as Job Creators: A Survey and Interpretation of the Evidence." *Small Business Economics* 35 (2): 227–44.

IMF (International Monetary Fund). 2014. *Republic of Croatia: 2014 Article IV Consultation-Staff Report; Press Release; and Statement by the Executive Director for the Republic of Croatia*. IMF Country Report 14/124.

Iootty, M., P. Correa, S. Radas, and B. Škrinjarić. 2014. "Stylized Facts on Productivity Growth: Evidence from Firm Level Data in Croatia: Vol. 1." Policy Research Working Paper WPS 6990, World Bank, Washington, DC.

Leighton, D. S. R. 1970. "The Internationalization of American Business: The Third Industrial Revolution." *Journal of Marketing* 34 (3): 3–6.

Levinsohn, J., and A. Petrin. 2003. "Estimating Production Functions Using Inputs to Control for Unobservables." *Review of Economic Studies* 70 (2): 317–42.

Nestić, D., A. Mervar, I. Rubil, B. Škrinjarić, M. Tkalec, and I. Tomić. 2014. "Recent Developments; Policy Assumptions and Projections Summary; Uncertainties and Risks to Projections." *Croatian Economic Outlook Quarterly* 16 (57): 1–12.

OECD (Organisation for Economic Co-operation and Development). http://www.oecd .org/unitedstates.

———. "Entrepreneurship at a Glance 2013." and FINA databases.

World Bank. 2009. *Croatia EU Convergence Report: Reaching and Sustaining Higher Rates of Economic Growth*. Report 48879-HR. Washington, DC: World Bank.

World Bank Group Entrepreneurship Snapshots. http://www.doingbusiness.org/data /exploretopics/entrepreneurship.

World Development Indicators database. http://data.worldbank.org/data-catalog/world -development-indicators.

World Economic Forum. 2013. *The Global Competitiveness Report 2013–2014*. Geneva: World Economic Forum. http://www.weforum.org/gcr.

CHAPTER 5

Innovation Challenges for Smart Specialization

Chapter Summary[1]

Better innovation performance would help close the productivity gap with Croatia's EU competitors. The contribution of innovation to sales growth, labor productivity growth, and total factor productivity (TFP) is lower than that of its peers (and see chapter 4). The country also underperforms on the contribution of R&D expenditures per worker to firm performance, though returns to R&D are higher than returns to infrastructure or education.

Croatia shows lackluster performance in business research and innovation. Despite the economic benefits that innovation can have on firms—whether in competitiveness, sales expansion, employment growth, or even survival—the private sector shows only moderate performance on this metric. It is primarily the country's large and old companies that innovate, and there is a "missing middle," as medium firms invest little in R&D. This problem stems mainly from limited access to internal and external resources (funds and qualified personnel), as well as market factors, including unfair competition and uncertain demand. Evidence on the composition of innovative activities shows that companies engage primarily in quality upgrading but do not perform well on introducing new goods and services. The exit rates in both activities are high relative to peers such as Slovenia.

Factors holding back innovation include the tax regime, lack of early-stage financing, and the business environment. One structural problem is that business R&D is low, despite generous tax breaks; and although small firms are the majority of beneficiaries, large firms receive most of the benefits. Sector concentration, too, is unfavorable, as only a few sectors account for the majority of these tax breaks. A second factor, the lack of early-stage finance, risks the premature death of potentially viable innovative start-ups. There is little supply of venture capital, which, combined with apparently substantial demand, creates a wide financing gap. Finally, Croatia has an unfriendly business environment, per the Doing Business indicators.

Research excellence and science–industry collaboration are lower than the EU average. The quality of publications is low, both when scientific publications feature among the top 10 percent most cited worldwide and in average citation impact. The country also falls slightly below the EU average in the share of doctorate graduates. Innovation capacity building is also limited by the lack of links between research institutes and private industry, feeding through into poor performance in the number of public–private copublications.

Perhaps the biggest challenge for boosting research and innovation impact is strengthening policy governance. The system is not fully functional: public funds are allocated without clear prioritization and results orientation. Having one single organization defining research and innovation programs simply because it is the managerial authority may generate highly inefficient allocation if this is not preceded by a coordination body. The country also needs to have research systems that are competitive and transparent, with quality-driven recruitment practices and efficient administrative procedures.

What Is the Importance of Research and Innovation?

Innovation is crucial to increasing total factor productivity (TFP), which is vital because differences in GDP growth across countries are mainly explained by differences in TFP. Further, innovation is one of the most important sources of firm growth. New inventions allow companies to increase production without having to hire more employees or to rent more capital, as process innovation reduces production costs and helps firms develop a comparative advantage. Organizational innovation makes companies more efficient, enhancing their competitiveness in domestic and international markets. Moreover, innovation develops a firm's ability to identify, assimilate, and exploit knowledge from the environment, which is essential for growth (Cohen and Levinthal 1990). Further, product innovation may lead to employment growth (Mohnen and Hall 2013).

For Croatia, higher R&D is likely to raise GDP and export levels, according to a 2009 World Bank study. The study compares the impact of five Lisbon Agenda targets on GDP and exports in Croatia and other EU countries, showing that increasing aggregate R&D to 3 percent of GDP (with 2 percent coming from the private sector, the Lisbon target for R&D) would raise GDP by 5.8 percent and exports by 13 percent above their baseline by 2025. Among six comparator countries, the impact of the R&D target on exports is for Croatia second only to that of Romania (table 5.1). Although this study assumes that all countries will reach their targets, and that the government fully controls business R&D (assumptions that require caution), it illustrates the potentially significant impact of R&D expenditures for Croatia.

Estimates of social rates of return to R&D in Croatia compare favorably with other areas for public investment. Seker (2011) shows that estimated rates of return to R&D in Croatia (73 percent) are at least double the value of returns to

Table 5.1 Impact of the Lisbon Agenda R&D Targets on GDP and Exports

	GDP	Exports
Romania	11.7	13.5
Croatia	5.8	12.9
Slovenia	6.9	10.5
Slovak Republic	8.9	10.4
Poland	5.5	8.5
Bulgaria	13.1	8.3
Hungary	6.4	8.0

Source: World Bank 2009.

Table 5.2 Impact of Innovation on Productivity

Independent variable	Sales growth	Labor productivity growth	TFP	Sales growth	Labor productivity growth	TFP
	(Column 1)	(Column 2)	(Column 3)	(Column 4)	(Column 5)	(Column 6)
Innovate	0.092	0.063	0.047	0.056	0.044	0.026
	(0.037)**	(0.033)*	(0.095)	(0.031)	(0.032)	−0.081
Innovate*Croatia	−0.072	−0.096	−0.329	−0.081	−0.083	−0.278
	(0.038)*	(0.035)**	(0.104)***	(0.033)**	(0.037)*	(0.081)**

Source: Seker 2011.
Note: All the regressions control for age, access to finance, exporter status, foreign ownership, industry fixed effects, and country fixed effects. Columns 1, 2, 4, and 5 also control for the lag of the dependent variable. Columns 3 and 6 control for the lag of sales. Columns 4, 5, and 6 control for training of permanent full-time employees, average duration of power outages, and interaction terms of these variables with a dummy variable for Croatia. That dummy variable takes the value of 1 if the country is Croatia. Standard errors are clustered by country. TFP is calculated as a residual of labor, capital, and raw materials.
Significance level: * = 0.10, ** = 0.05, *** = 0.01.

infrastructure (around 24–34 percent) and seven times as high as those to education (around 10 percent)—or 10 times as high if the rate of return for schooling for high-income countries (7.4 percent) is used. Higher returns for R&D in Croatia are consistent with the country's lower stock of R&D capital than its stock of infrastructure and human capital.

Better innovation performance would help close the productivity gap between enterprise sectors in Croatia and EU competitors. Seker (2011) analyzed the effect of innovation on productivity using World Bank Enterprise Surveys (2005–09), and showed that the contribution of innovation (creation of a new product or service, or upgrading of an existing product line or service) to sales growth is 7.2 percent less than the EU-8 average (table 5.2).[2] A similar situation obtains for labor productivity growth, which is 9.6 percent lower in Croatia than the EU-8 average. The difference is remarkably high (32.9 percent) for TFP. The results are robust with controlling other variables such as skills and infrastructure (columns 4, 5, and 6).

Table 5.3 Impact of R&D on Productivity

Independent variable	Sales growth	Labor productivity growth	TFP
	(Column 1)	(Column 2)	(Column 3)
Log(R&D/workers)	0.063	0.107	0.096
	(0.042)	(0.044)**	(0.041)**
Log(R&D/ workers)*Croatia	−0.061	−0.046	−0.088
	(0.024)**	(0.016)**	(0.027)**

Source: Seker 2011.
Note: All the regressions control for age, access to finance, exporter status, foreign ownership, industry fixed effects, and country fixed effects. Columns 1 and 2 also control for the lag of the dependent variable. Column 3 controls for the lag of sales. "Croatia" is a dummy variable that takes the value of 1 if the country is Croatia. Standard errors are clustered by country. TFP is calculated as a residual of labor and capital.
Significance level: * = 0.10, ** = 0.05, *** = 0.01.

Croatia also underperforms its European peers on the contribution of R&D per worker on firm productivity (table 5.3). In line with the previous analysis, this finding shows that Croatia's R&D contribution to sales growth is 6.1 percent less than the EU-8 average. The difference is 4.6 percent for labor productivity and 8.8 percent for TFP.

Recent Innovation and R&D Performance

Business Innovation

On the basis of indicators from the European Commission's *Innovation Union Scoreboard 2014*, Croatia ranks moderately among EU members. It ranks 22nd of 30 countries on the share of innovative firms (42 percent) out of total enterprises participating in the survey (figure 5.1). It performs better on the share of companies conducting in-house R&D (ranked 11; figure 5.2) and external R&D (ranked 13; figure 5.3).[3]

Large companies appear more innovative than small and medium enterprises (SMEs). Some 40 percent of Croatia's small firms (10–49 employees) innovate (figure 5.4), and the country ranks 19 on this indicator, close to Spain and Norway. The share of SMEs introducing marketing or organizational innovations is below the EU regional average, however, having slipped by 4.3 percent over 2000–11. The share of SMEs conducting product or process innovation is also below the EU average, although it edged up by 1.8 percent over the period. For medium companies (20–249 employees), the country ranks 21 (figure 5.5). Among large firms (250 or more employees), more than 70 percent innovate (figure 5.6), but even then the country ranks 19, behind peers like Bulgaria, the Czech Republic, Estonia, and Slovenia.

Thus, medium companies do not innovate much, for several reasons. According to the 2010 Community Innovation Survey, limited access to internal and external resources (funds and qualified personnel) as well as market factors, including unfair competition and uncertain demand, are regarded as the major

Figure 5.1 Share of Innovative Firms, 2010

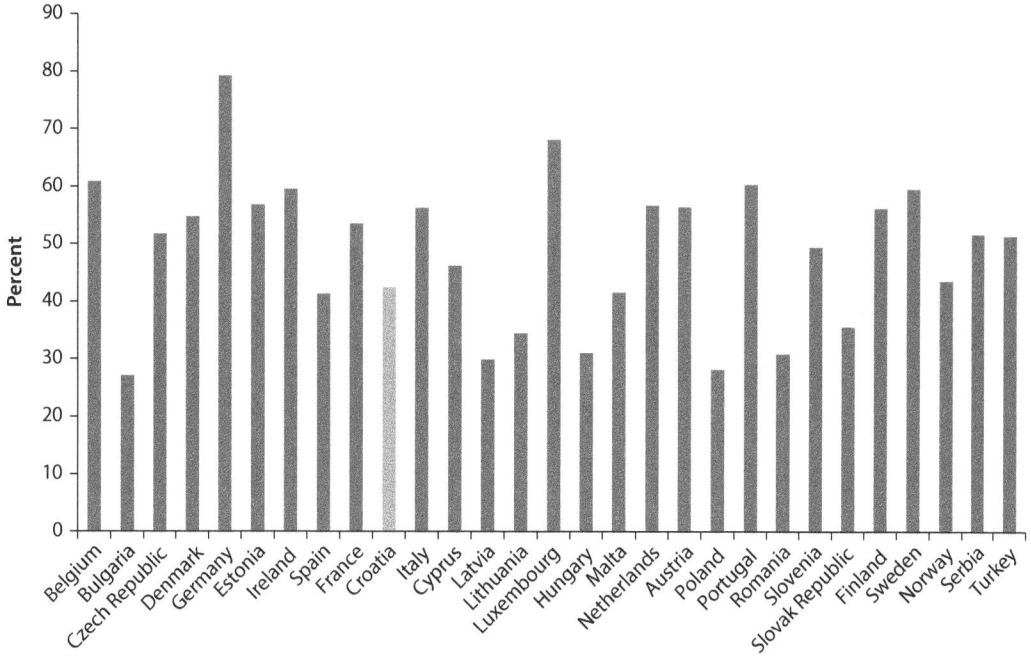

Source: Community Innovation Survey 2010.

Figure 5.2 Share of Innovative Firms Conducting in-House R&D, 2010

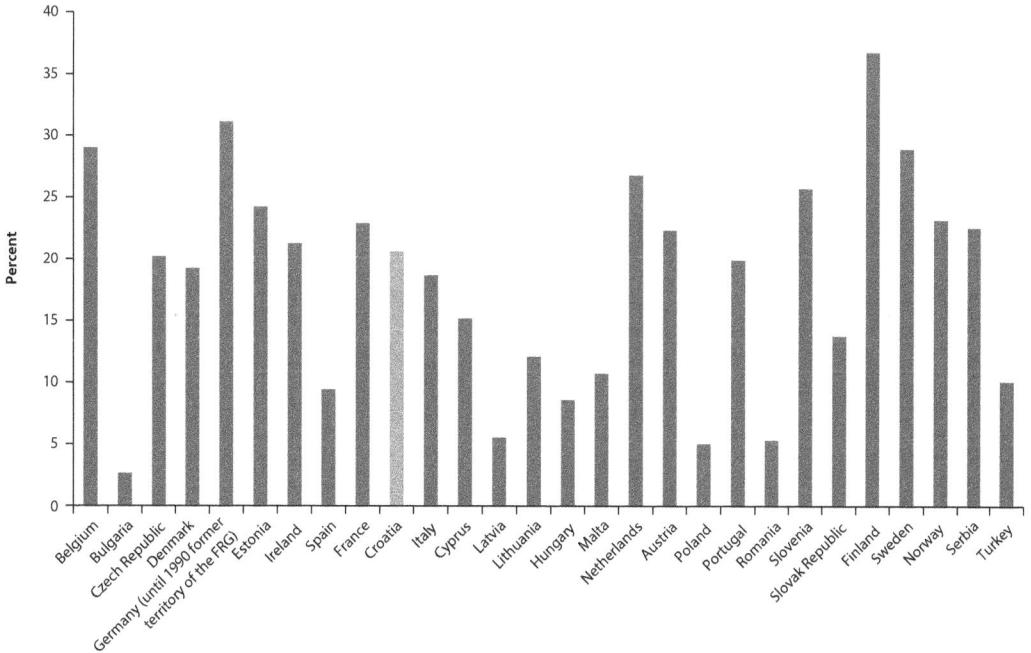

Source: Community Innovation Survey 2010.

Smart Specialization in Croatia • http://dx.doi.org/10.1596/978-1-4648-0458-8

Figure 5.3 Share of Innovative Firms Conducting External R&D, 2010

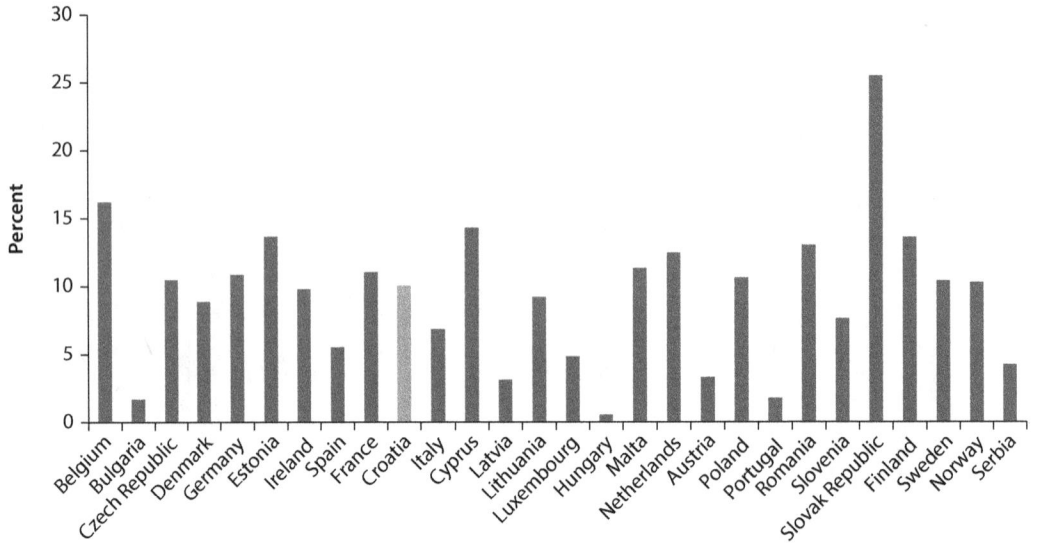

Source: Community Innovation Survey 2010.

Figure 5.4 Share of Small, Innovative Firms, 2010

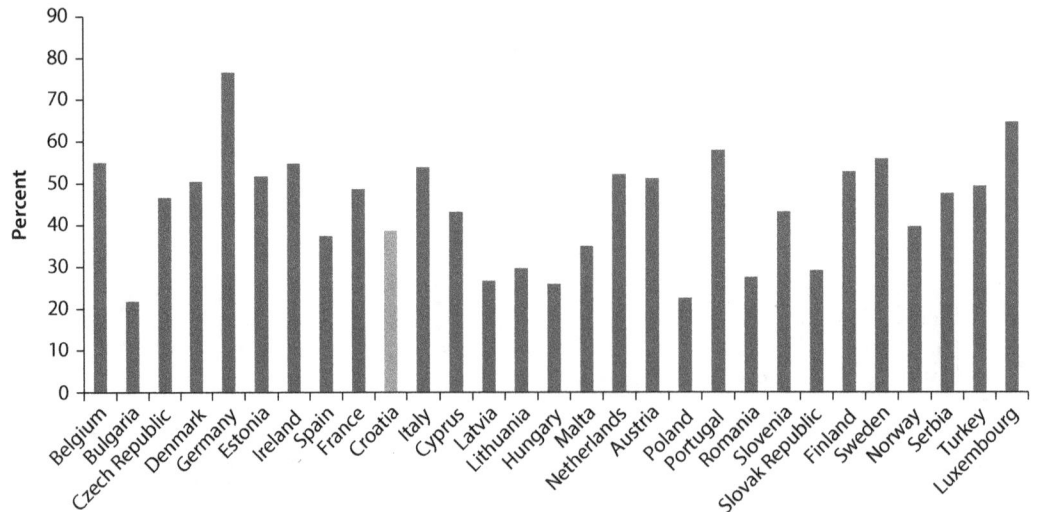

Source: Community Innovation Survey 2010.

barriers by at least 25 percent of respondents. These findings come through in other studies: Radas and Božić (2009), for example, find that financing and innovation costs are the most important hurdle in the way of Croatian companies innovating, followed by lack of qualified employees and limited information on technology and markets. OECD (2011) shows that lack of qualified personnel

Figure 5.5 Share of Medium, Innovative Firms, 2010

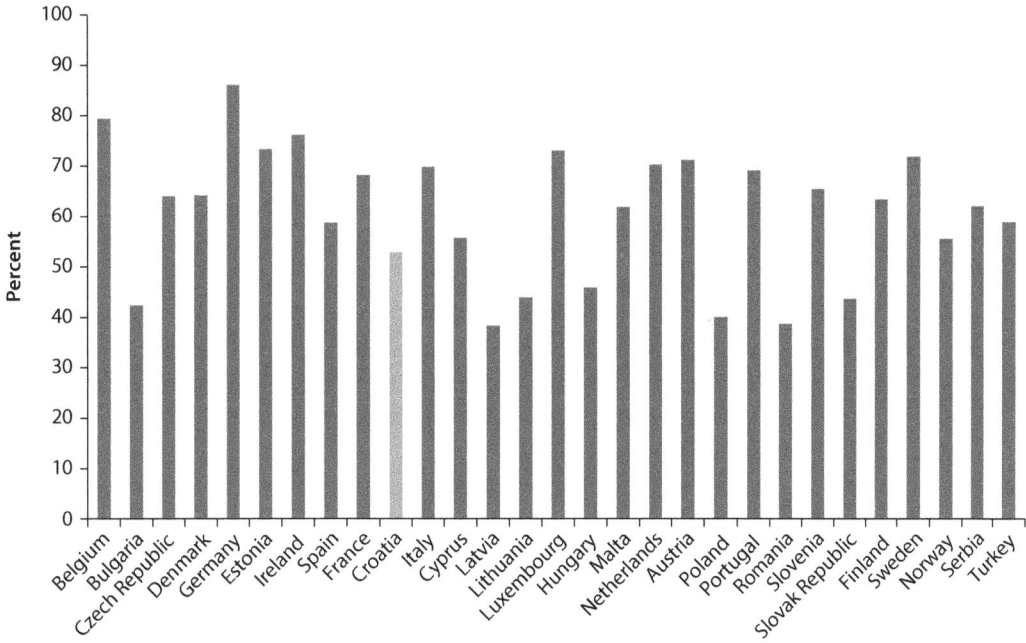

Source: Community Innovation Survey 2010.

Figure 5.6 Share of Large, Innovative Firms, 2010

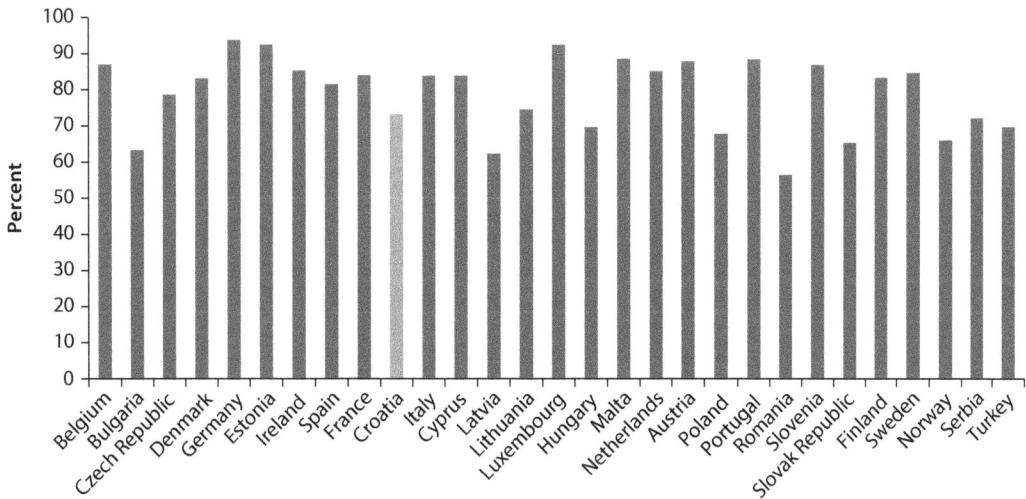

Source: Community Innovation Survey 2010.

ranked as the second most important barrier to innovation, although only 44 percent of Croatian firms said that they faced difficulty in finding skilled employees. Also according to this study, employees frequently lack experience (80 percent) or the education system does not equip them with the right set of skills (40 percent).

Composition of Innovation

Vertical innovation is more common than horizontal innovation in Croatia. World Bank Enterprise Surveys show that firms upgrade the quality of an existing product or service (vertical innovation) more than they create a new product or service (horizontal innovation) (figure 5.7). Over 2005–09, 58 percent of firms innovated vertically, but only 22 percent did so horizontally. There is more entry (No–Yes) in the horizontal activity (36 percent), partly because most of the companies conduct vertical innovation. The exit rates (Yes–No) are similar for both types, around 20 percent.

Croatia's pattern has some similarities to Slovenia's pattern, although Slovenia shows no exit activity for horizontal innovation (figure 5.8).[4] In Slovenia, the intensive margin[5] is predominately associated with vertical innovation, as 40 percent of firms innovated vertically in 2005 and 2009, but only 25.71 percent innovated horizontally. Among the firms, 20 percent never conducted horizontal innovation, versus 5.71 percent never innovating vertically. The entry margin is similar for vertical and horizontal innovations.

Figure 5.7 Percentage of Firms Innovating, Croatia, 2005–09

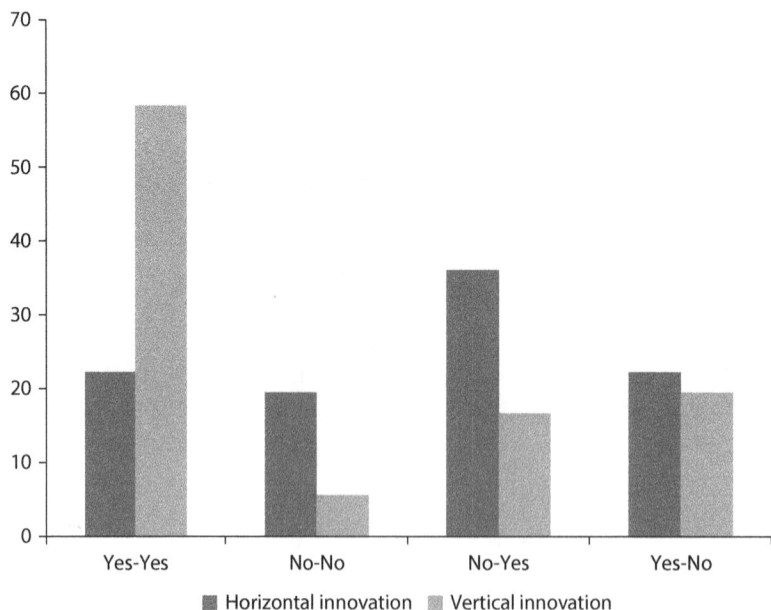

Source: World Bank Enterprise Surveys 2005–09.

Figure 5.8 Percentage of Firms Innovating, Slovenia, 2005–09

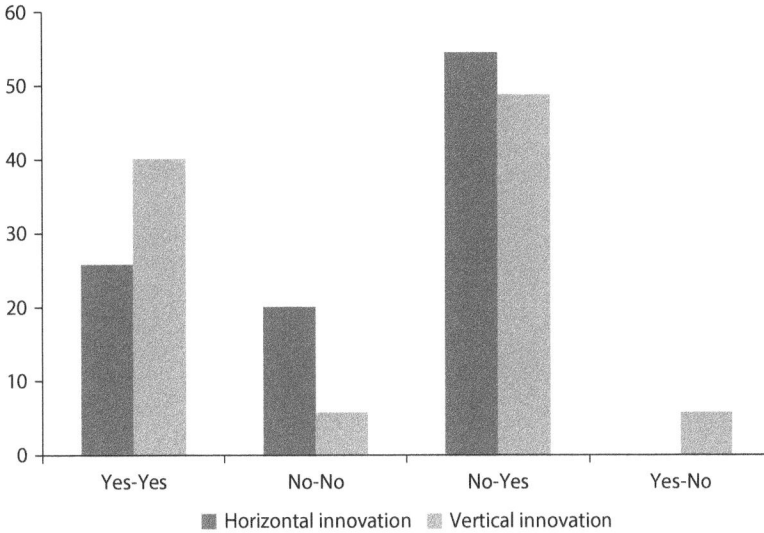

Source: World Bank Enterprise Surveys 2005–09.

Croatia's SMEs perform much better than most peers on introducing products and process innovations (Figure 5.9). Its share of SMEs is not too far from the EU-27 average (the same applies to SMEs introducing marketing or organizational innovations as a share of total SMEs). However, Croatia's performance on business enterprise R&D (BERD)—and business enterprise researchers (full-time equivalent) per 1,000 labor force—is close to or even below that of the reference group and much lower than the EU-27 average. In 2000–11, "non-R&D related expenditures" grew more than six times, to become 8 percent higher than the EU average. (European Commission, 2013)

Business Investment in R&D

Croatian enterprises are too little involved in R&D. The country displays a low level of business enterprise R&D (BERD) as a share of GDP against the EU-28 average, having underperformed since the middle of the 2000s versus both that average and versus Slovenia, one of its main comparators (on share of business expenditure on R&D as a share of total R&D spending) (figure 5.10). And although both Slovenia and the EU average increased per capita BERD over 2006–13, this indicator stagnated in Croatia for the most part of the period (figure 5.11).

Business investment in R&D is concentrated in a few multinational companies. Most of these firms are in the pharmaceutical, telecommunications, agricultural, and food and beverage industries, and include Galapagos (formerly GlaxoSmithKline Research Centre); PLIVA Institute

Figure 5.9 Key Innovation Indicators

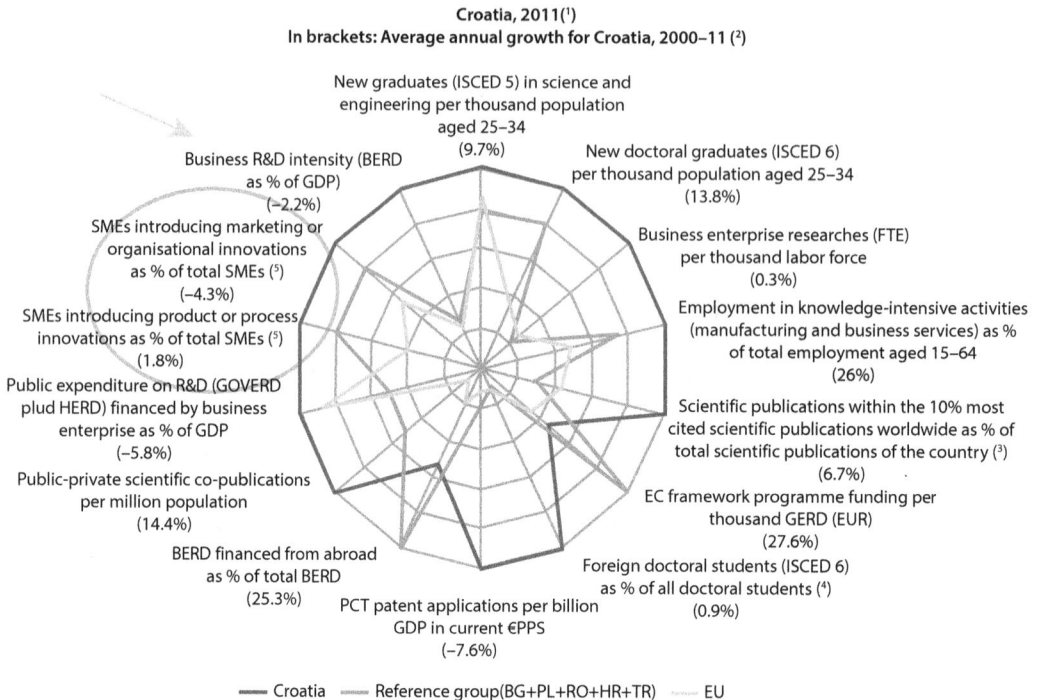

Croatia, 2011([1])
In brackets: Average annual growth for Croatia, 2000–11 ([2])

New graduates (ISCED 5) in science and
engineering per thousand population
aged 25–34
(9.7%)

New doctoral graduates (ISCED 6)
per thousand population aged 25–34
(13.8%)

Business R&D intensity (BERD
as % of GDP)
(−2.2%)

SMEs introducing marketing or
organisational innovations
as % of total SMEs ([5])
(−4.3%)

Business enterprise researches (FTE)
per thousand labor force
(0.3%)

SMEs introducing product or process
innovations as % of total SMEs ([5])
(1.8%)

Employment in knowledge-intensive activities
(manufacturing and business services) as %
of total employment aged 15–64
(26%)

Public expenditure on R&D (GOVERD
plud HERD) financed by business
enterprise as % of GDP
(−5.8%)

Scientific publications within the 10% most
cited scientific publications worldwide as % of
total scientific publications of the country ([3])
(6.7%)

Public-private scientific co-publications
per million population
(14.4%)

EC framework programme funding per
thousand GERD (EUR)
(27.6%)

BERD financed from abroad
as % of total BERD
(25.3%)

Foreign doctoral students (ISCED 6)
as % of all doctoral students ([4])
(0.9%)

PCT patent applications per billion
GDP in current €PPS
(−7.6%)

—— Croatia   ····· Reference group(BG+PL+RO+HR+TR)   ···· EU

Source: European Commission. 2013. "Research and Innovation performance in EU Member States and Associated countries." Used with permission; further permission required for reuse.
Notes: (1) The values refer to 2011 or to the latest available year.
(2) Growth rates which do not refer to 2000–2011 refer to growth between the earliest available year and the latest available year for which comparable data are available over the period 2000–2011.
(3) Fractional counting method.
(4) EU does not include DE, IE, EL, LU, NL.
(5) TR is not included in the reference group.

(a pharmaceutical company); Ericsson-Nikola Tesla Institute (ICT); Podravka (food); Koncar-Electrotechnical Institution; Belupo (medicinal products); INA-Oil Company; and the Civil Engineering Institute of Croatia.

The R&D intensity (R&D spending as a share of turnover) of medium firms is very low compared with that of other countries in the region. Among Croatia's large, medium, and small firms, the R&D intensity of medium firms is the poorest relative to EU peers, at only 0.16 percent (ranked 21; figure 5.14). Overall R&D intensity is 1.17 percent (ranked 6; figure 5.12), putting the country above the Netherlands, among others; among small firms it is 0.34 percent (ranked 16; figure 5.13); and among large firms it is 1.98 percent (ranked 4; figure 5.15).

The low R&D intensity of medium firms is also confirmed by firm-level data from Enterprise Surveys.[6] Figure 5.16 presents a box-plot representation of R&D intensity in 2009. The box displays three percentiles: 25, 50, and 75. (The same graph for Slovenia is also presented; figure 5.17.) One striking pattern is the lack of innovation activity by medium firms, because the median R&D intensity of these firms is equal to zero, pointing to a "missing

Figure 5.10 BERD (% of GDP)

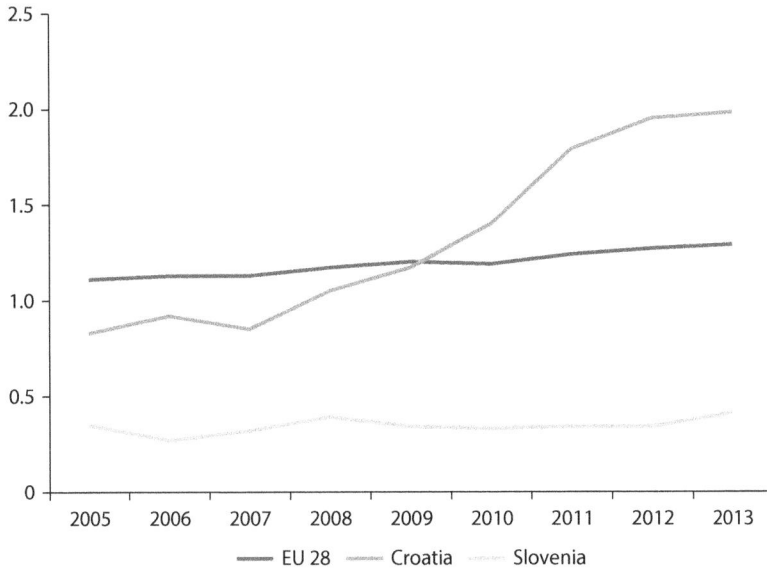

Source: Eurostat.

Figure 5.11 BERD (€ per Capita)

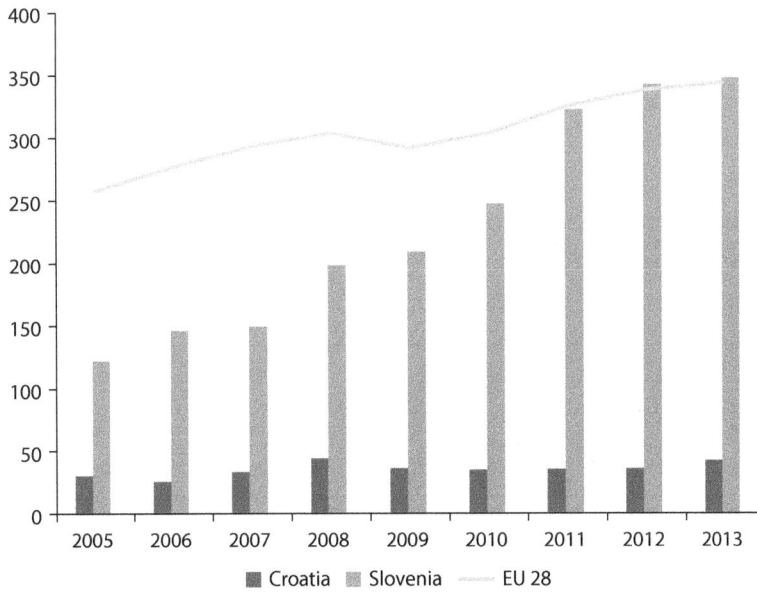

Source: Eurostat.

Figure 5.12 R&D Intensity (All Firms), 2010

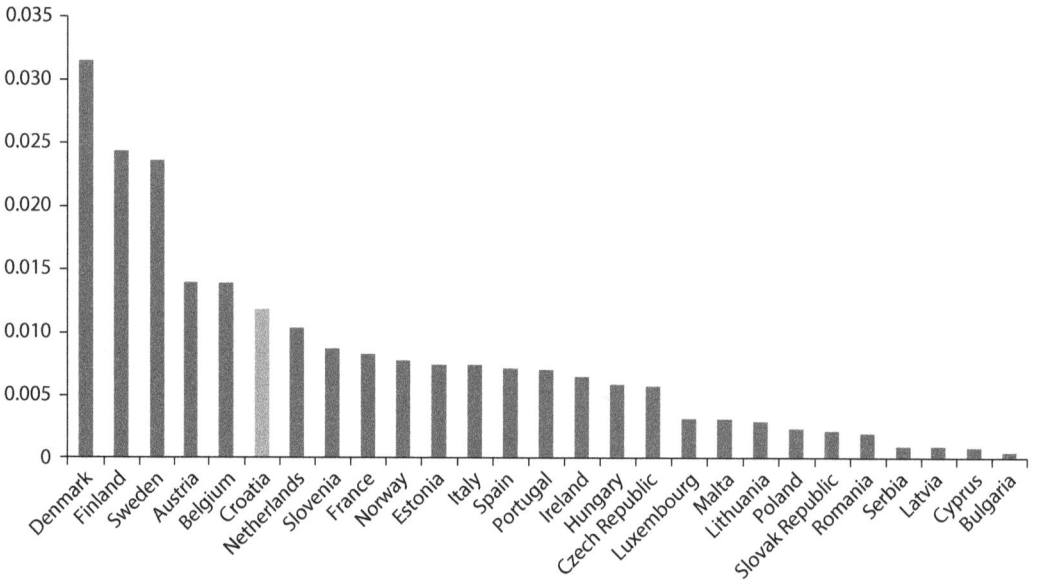

Source: Community Innovation Survey 2010.

Figure 5.13 R&D Intensity (Small Firms), 2010

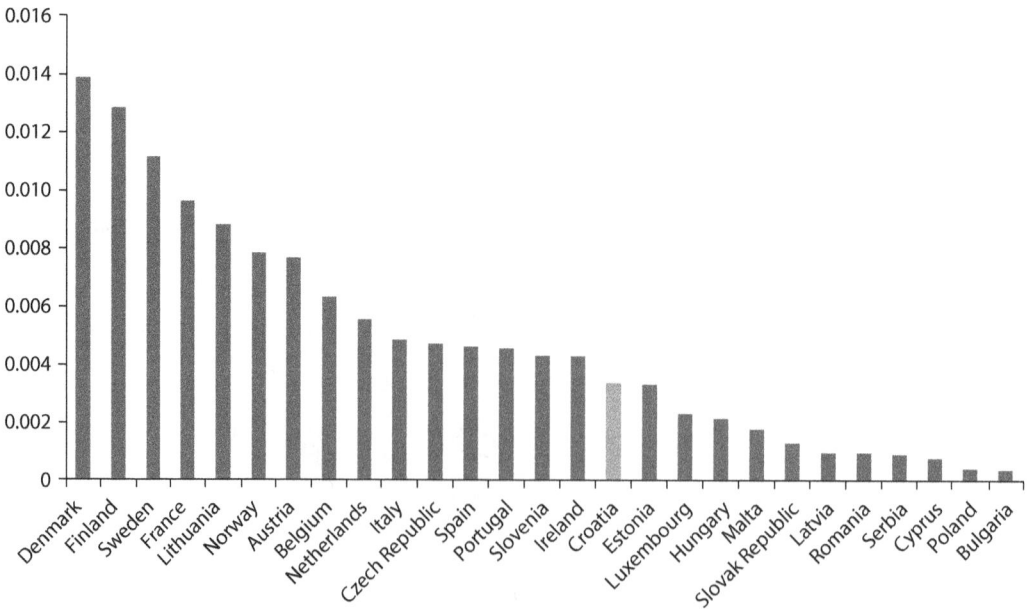

Source: Community Innovation Survey 2010.

Figure 5.14 R&D Intensity (Medium Firms), 2010

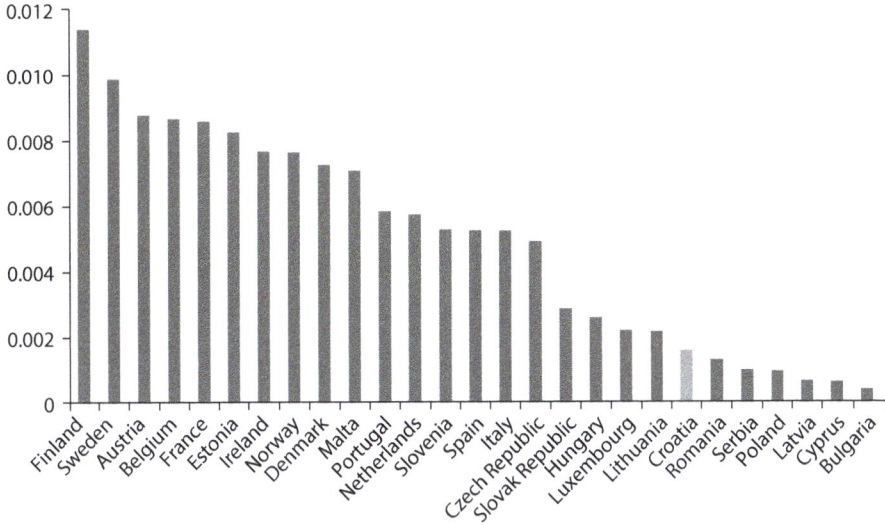

Source: Community Innovation Survey 2010.

Figure 5.15 R&D Intensity (Large Firms), 2010

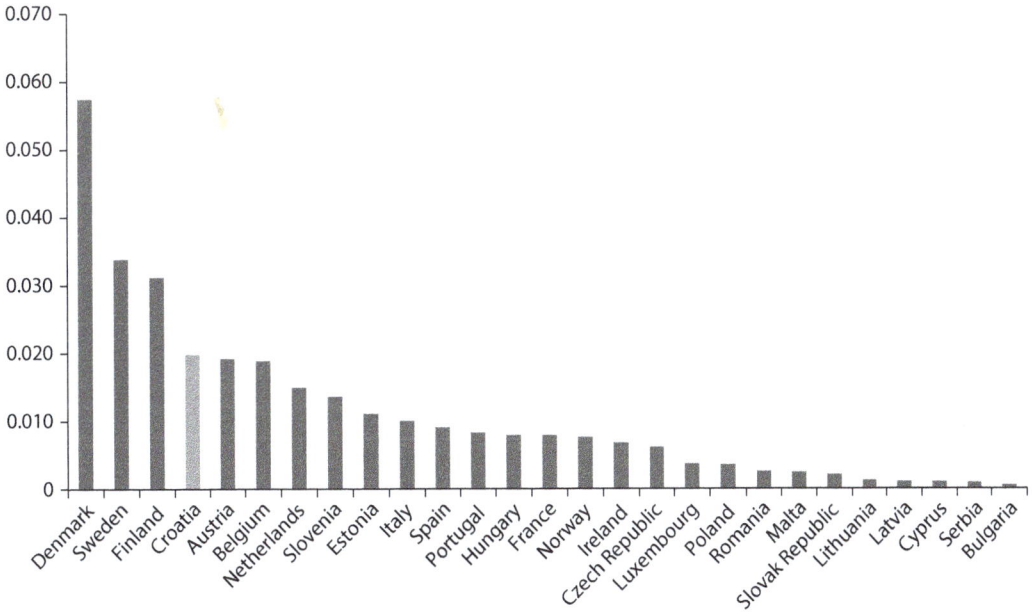

Source: Community Innovation Survey 2010.

Figure 5.16 R&D Intensity (by Size), Croatia, 2009

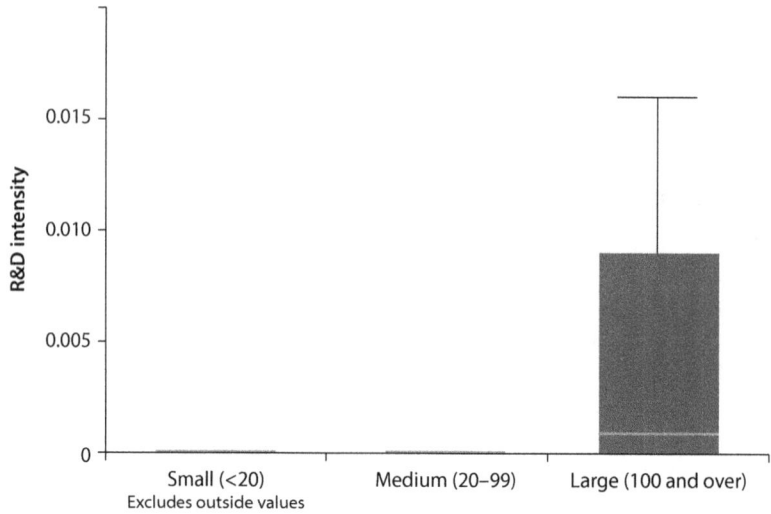

Source: World Bank Enterprise Surveys 2005–09.

Figure 5.17 R&D Intensity (by Size), Slovenia, 2009

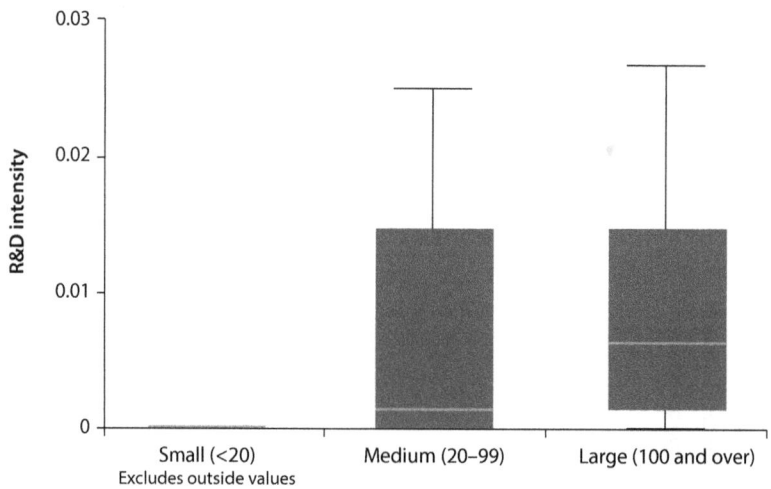

Source: World Bank Enterprise Surveys 2005–09.

middle" in Croatia.[7] (The missing middle is also linked to the overall lack of dynamism of local firms, as presented in chapter 4.) Also, old (16+ years) companies are the only ones conducting R&D (figure 5.18), unlike in Slovenia, where young (1–5 years) and medium-age (6–15 years) firms also carry out R&D (figure 5.19).

Figure 5.18 R&D Intensity (by Age), Croatia, 2009

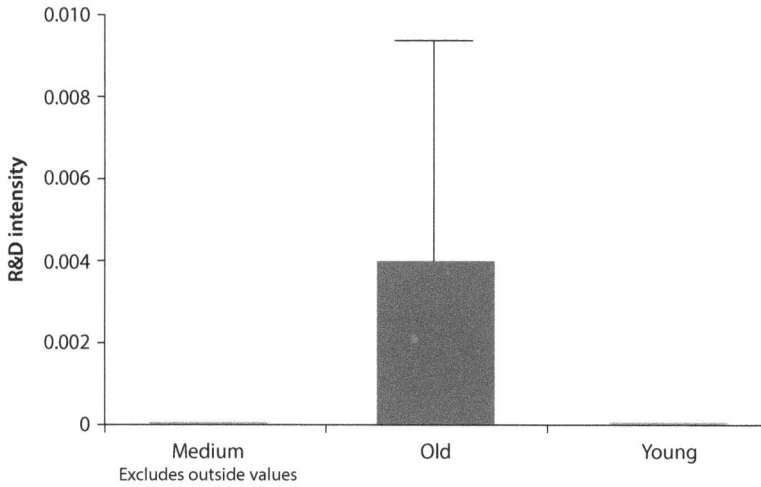

Source: World Bank Enterprise Surveys 2005–09.

Figure 5.19 R&D Intensity (by Age), Slovenia, 2009

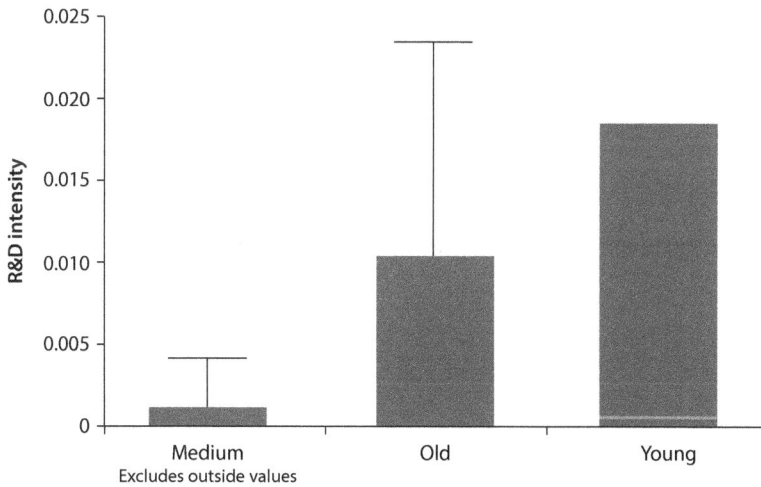

Source: World Bank Enterprise Surveys 2005–09.

Business R&D Determinants

Several factors contribute to Croatia's poor performance on R&D investment by businesses: concentration of tax incentives by firm size and sector; lack of early-stage financing; and the poor business environment. Modest research excellence and barriers to science–industry collaboration, as well as pervasive weaknesses in the governance framework—perhaps the biggest challenge—also loom large.

Concentration of Tax Breaks among Large Firms and in Two Industrial Sectors

Although small firms form the majority of the beneficiaries of R&D tax incentives, large firms receive most of the benefits; tax breaks are also highly concentrated by sector. Croatia's system of tax incentives for R&D is in fact one of the most generous in the OECD, yet business R&D stays low. The country provided in 2008 a tax break of about 35 percent for US$1 of R&D, second only to France (42 percent)[8] (figure 5.20). A recent study of R&D tax incentives (Švaljek 2012) shows that around 77 percent of state aid for R&D came from tax incentives in 2009 (€14.6 million). Despite a change to the tax regime in 2007, the tax advantage for R&D as a share of GDP did not change much, as the average for 2007–09 was 0.05, against 0.06 in 2004–06. There were 272 beneficiaries of the R&D tax incentive in 2008 and 261 in 2009. The incentives were so high that 27 percent of beneficiary companies fully eliminated their tax liability in 2009. According to the Švaljek study, the statutory profit tax rate in Croatia is 20 percent, but companies using R&D incentives had to pay an average of only 6.5 percent in 2008 and 12.4 percent in 2009.

Small firms represented the majority of the R&D tax-break beneficiaries in 2009 (figure 5.21), but large firms received most of the benefits (figure 5.22). Indeed, Švaljek (2012) found a positive correlation between total revenues and R&D tax incentives. The study shows that in 2008, 90 percent of all tax incentives were attributed to only nine companies, or 3.3 percent of the recipients. In 2009, 27 companies (10.3 percent) got 90 percent of the total, a slightly wider distribution of benefits. Small firms have the highest rate of state aid as

Figure 5.20 R&D Tax Breaks, Share of Represented Tax Breaks of Total Funding

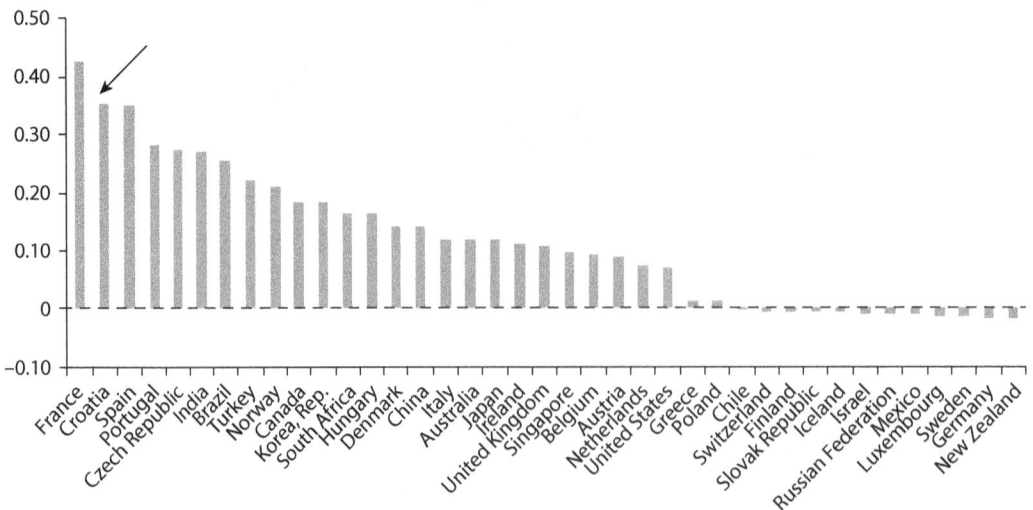

Source: Correa and Hansen 2010.

Figure 5.21 Number of Beneficiaries, 2009

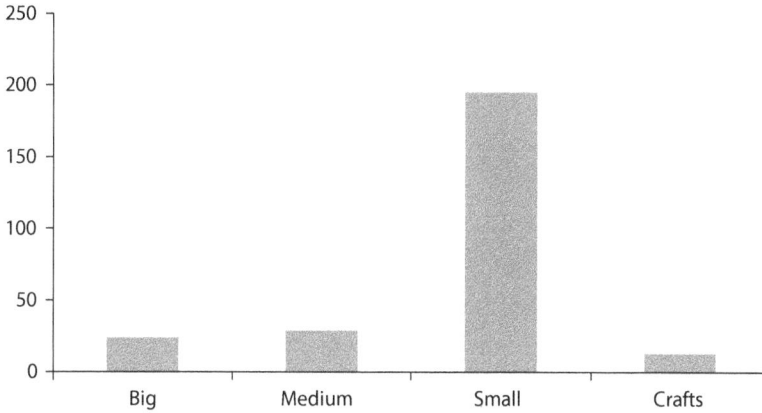

Source: Švaljek 2012.

Figure 5.22 Share in Total Net Profit, 2009
Percent

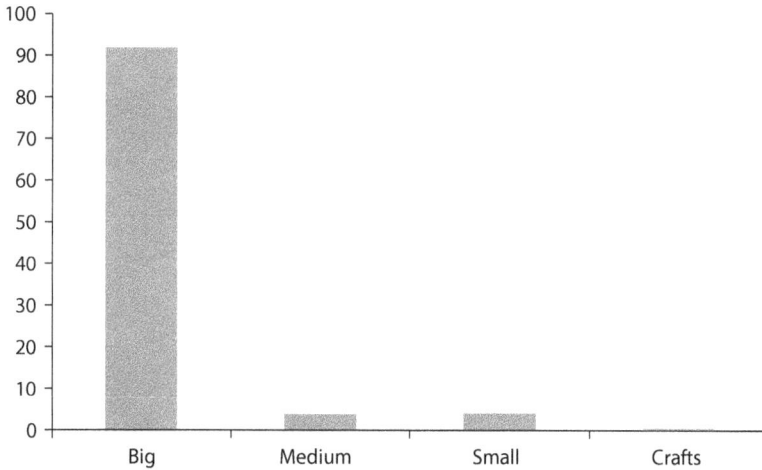

Source: Švaljek 2012.

a percentage of net profits (figure 5.23). The average tax burden (the tax liability after tax deductions in net profit) was the highest for crafts (figure 5.24).

Tax breaks are also highly concentrated by sector. Two industrial sectors—manufacture of radio, television, and communication equipment and apparatus; and manufacture of chemicals, chemical products, and man-made fibers—accounted for 77.8 percent of R&D tax incentives in 2008 and 62.1 percent in 2009. Nearly all of the benefits accrued to the city of Zagreb (93.7 percent).

Figure 5.23 State Aid Based on R&D Incentives, 2009
Percent of net profit

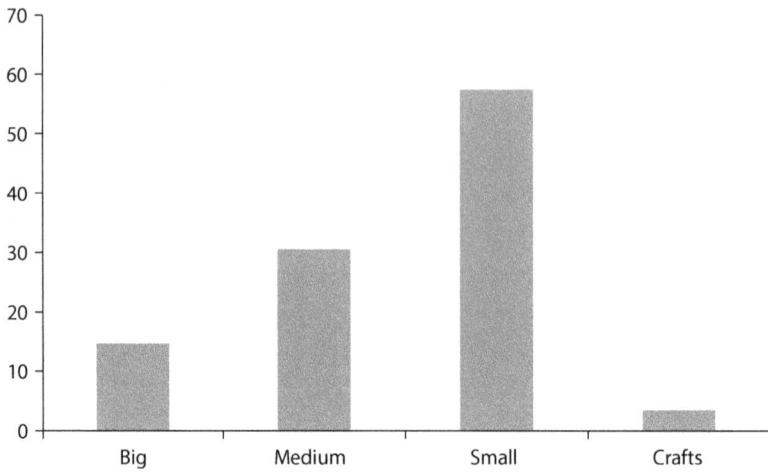

Source: Švaljek 2012.

Figure 5.24 Average Tax Burden (% of Tax Liability after Tax Deductions in Net Profit), 2009

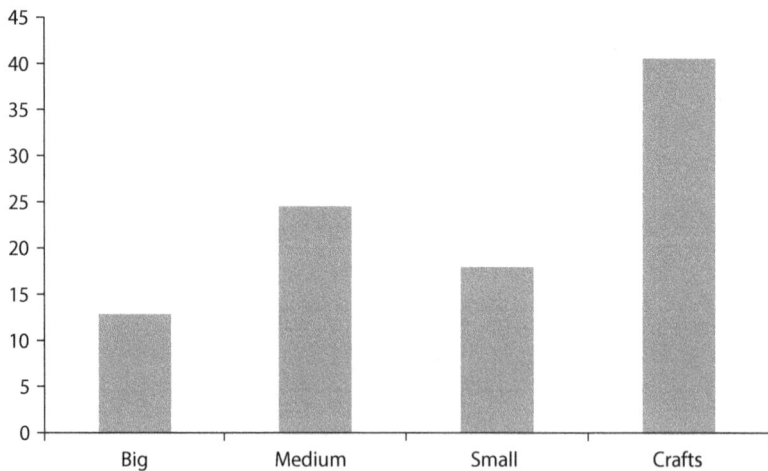

Source: Švaljek 2012.

A significant number of potential beneficiaries refrain from using the incentives because of concerns over excessive red tape and risks of corruption.

The Financing Gap for Venture Capital[9]

The lack of early-stage finance (first and second round investment) carries the risk of premature death in the case of potentially viable innovative start-ups. Access to such financing is key to unlocking their potential, which tends to be very innovative, focusing on high-tech fields such as biotechnology, clean energy, and software. Export-orientated start-ups are in a position to leverage the country's low cost base to sustain long periods of high export growth.

There seems to be healthy demand for early-stage financing: several government programs have, for example, supported the development of technologies with commercial potential. The Unity through Knowledge Fund (UKF) has financed scientific research, including through programs focused on commercializing new technologies. One, the Proof of Concept initiative, supports entrepreneurs' precommercial activities by helping them develop new products, services, and technological processes. Similarly, the IRCRO Research and Development Program aims to enhance cooperation on R&D between SMEs and scientific and research institutes (which is badly needed—see below). Programs supporting start-ups include the Technology Infrastructure Development Programme (TEHCRO), which focused on developing technology incubators and technology and business centers. By the end of 2010, 52 companies were residents of the TEHCRO incubators.[10] These last three initiatives were financed by the Business Innovation Croatian Agency (BICRO), which has in all financed seven programs, which in turn supported more than 250 projects with about €40 million (table 5.4).[11]

UKF can be highlighted as a success. It aimed to promote research excellence through collaboration between Croatia's scientific diaspora and local researchers with the competitive provision of research grants.[12] In 2007–11, 325 project applications were submitted (for €24.8 million) of which 91 were funded (€7.8 million), benefiting 544 researchers from 260 institutions. Third parties (including foreign research institutes and SMEs) contributed with about 35 percent of the total investment. UKF funding contributed to the publication of 177 articles in peer-reviewed publications, including *Nature* and *Science*; and the establishment or development of collaboration with top-level research organizations, such as Yale and Stanford Universities and Harvard Medical School (United States), the Federal Institute of Technology, Lausanne (Switzerland),

Table 5.4 Support for Commercial Technologies and Start-Ups by HAMAG-BICRO, 2007–12

Program	Projects supported	Contracted amount (€ million)
POC	112	3.0
RAZUM	24	15.0
IRCRO	24	2.0
TEST	31	6.0
EUREKA	10	1.0
KONCRO	42	0.3
TECHRO	14	10.0
Total	**257**	**37.3**

Source: Own elaboration based on data from HAMAG-BICRO data.
Note: KONCRO and TEST are inactive.

the Royal Institute of Technology (Sweden), and the Max Planck Institute (Germany). Other evidence corroborates the idea that UKF had a positive impact on research excellence. First, the rate of approval of UKF-related applications to the EU Seventh Framework Program (about 30 percent) was about twice the Croatian average (15 percent). An assessment by the Institute of Economics of Zagreb (IEZ 2011) showed that the program, too, helped deepen collaboration (box 5.1). The program was also recognized as having best practices by the International Labour Organization and the European Commission.

The Proof of Concept and RAZUM programs, under a preliminary series of assessments (World Bank 2012, for example), had a positive impact on early-stage innovation. The Proof of Concept program was designed to support established and start-up businesses and universities spinning off companies in Croatia to develop new technologies, products, and business ideas. It has given innovative companies and researchers the opportunity to verify and validate technical properties, commercial viability, and the potential and feasibility of a research result by providing competitive grants to support external expenditures on precommercialization proof-of-concept activities. The RAZUM program, implemented by BICRO in 2007–12 to support SME investments in R&D (box 5.2), is a soft-loan mechanism to encourage the private sector to spend more on R&D and reduce the risk that firms often face in innovation processes (early precommercial stage).

Croatia has significant investable demand for venture capital funds—around €25 million a year for 2012–16, according to a study by the European Investment Fund; at least €16 million a year according to the Croatian Private Equity and

---

Box 5.1 Assessing the Impact of UKF

The assessment by the IEZ (2011) analyzed one of the subprograms, the "Across the Borders" grant. This is a collaborative research grant between Croatian scientists and scientists from abroad but of Croatian origin. Croatian scientists were interviewed as part of the assessment, including both beneficiaries and nonrecipients. The results showed the following:

Beneficiaries reported that they had applied to the UKF program primarily because of the need to finance their research. They said that without the UKF grant, it would have been far harder to continue their research, and the research output would have been greatly reduced.

Without the UKF grant, researchers would have found it much more difficult to establish connections with foreign scientists and those in the Croatian diaspora.

The main expected benefit for beneficiaries was access to new methods and better research results arising from closer collaboration. An overwhelming share of researchers surveyed (95 percent) declared that they would likely apply again for UKF funding.

A significant negative impact for nonbeneficiaries was more limited ability for young researchers and doctoral students to begin their careers at a higher level.

Source: Authors' elaboration based on IEZ (2011).

Box 5.2 Assessing the Impact of RAZUM

RAZUM's behavioral and output additionality was measured by a survey of beneficiaries. According to responses, RAZUM enabled beneficiary companies to increase their capacity for conducting innovation and R&D, and to extend staff knowledge and capability via the hiring of highly educated professionals. In most cases these changes are likely to be permanent. New product development was positively affected in a large majority of cases, suggesting better innovation capability. For most companies that received RAZUM support, work on the project generated additional ideas for innovations.

When interviewed on what would have happened had they not received the RAZUM grant, six companies (30 percent) reported that they would have abandoned the project entirely. The majority (86 percent) of the rest would have relied on their own resources, while some of them would have tried to obtain financing from banks and venture capital funds. Three firms would have attempted to identify sources of funding through strategic partner-ships and some other R&D subsidies. However, the absence of RAZUM money would not have been without consequences—most companies would have proceeded with the project on a smaller budget.

A similar pattern is found in the case of nonbeneficiaries. Those firms either were in the evaluation process (passed the preselection phase) or were approved and waiting for financing. In the hypothetical situation of not receiving RAZUM funding, two companies of 14 would have abandoned the planned project and started another one, whereas all other companies would have proceeded with their projects (no firm declared that it would not continue with that or any other project). However, the absence of RAZUM grant would have had consequences for the duration, scope, R&D capacity, and overall quality of projects:

Without the RAZUM grant, the vast majority of respondents said that they would have proceeded with the project but over a longer time frame (92.9 percent), on a smaller budget (85.7 percent), with reduced scope (85.7 percent), and with inadequate equipment and/or machinery (71.4 percent).

On outcomes, many companies would not have hired additional employees (71.4 percent), and the innovativeness level of the output would have been lower (42.9 percent).

Source: Authors' elaboration based on Radas et al. (2011).

Venture Capital Association; and €20 million according to Croatian technology transfer offices. The Croatian Business Angel Network calculates that around €2 million a year's worth of projects are not going ahead due to an absence of early-stage financing. This gap arises because Croatia has lower levels of venture capital investment than comparable countries in the region—perhaps one-tenth of that in the Czech Republic and one-third of that in Hungary, although Croatia is at a broadly similar level of development.

Venture capital varies sharply by year. The country does not benefit from any venture capital fund. Venture capital investment fluctuated widely over 2007–12, dropping close to zero in 2009, at the peak of the global crisis (figure 5.25).

Figure 5.25 Average Venture Capital Investment, 2007–12
€ Million

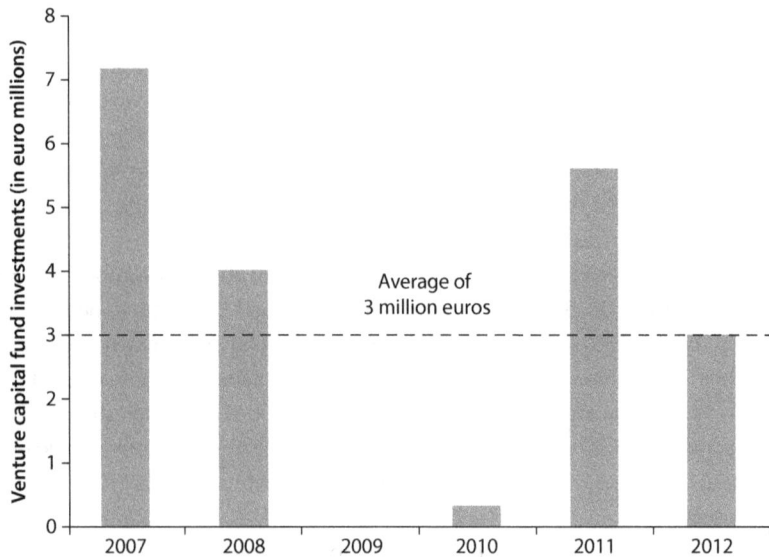

Source: Croatian Private Equity and Venture Capital Association.

Although this indicator has recovered somewhat, on average there has been only €3 million a year since 2007.

Some private equity funds have entered the market. In 2010, the government launched a tender to seed a domestic institutional private equity industry through the Economic Cooperation Funds program and committed 1 billion kunas (about US$184 million or €139 million) to match one-to-one financing raised by Croatian fund managers from private and institutional investors (domestic or international). In 2011, five funds attracted the necessary cofinance and began operating. The program contributed to a sharp increase in funding, from less than €100 million to more than €300 million (Krstulovic and Makek 2013).

However, private equity fund managers agree that those resources will not trickle down to support technology-driven start-ups. Croatian private equity funds are not suited to providing early-stage funding for several reasons, primarily the following:

- Early-stage investments are riskier than those typically made by private equity funds. These funds tend to seek the safety and capital preservation that comes from investing in more established companies with proven business models and management teams.
- Private equity funds invest over shorter time frames and plan to exit their investments after three to five years. This is shorter than the period needed for investment in start-ups, which tends to last from seven to 10 years.
- Private equity funds seek to make investments that are too large for start-ups.

Figure 5.26 Venture Capital Funding Gap, Croatia

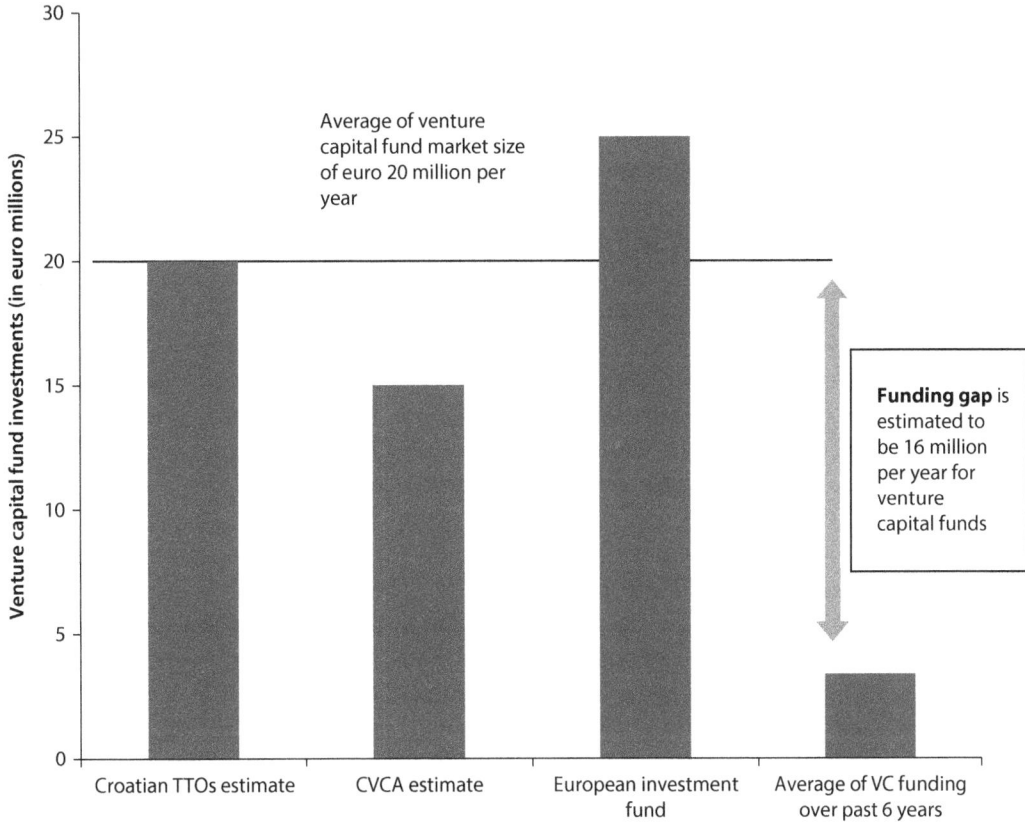

Source: Croatian Private Equity and Venture Capital Association.

The heavy demand for venture capital and the lack of supply point to a substantial early-stage financing gap. An average of estimates suggests annual investable demand of €20 million a year, but there is only €3 million in venture capital investment—an annual financing gap of roughly €17 million (figure 5.26). The financing gap has not therefore yet been properly addressed by the government or the private sector—potentially holding back the country's overall innovation and economic performance, and the government's ability to meet its commitments under the Europe 2020 Agenda.

Weaknesses in the Business Environment

Weaknesses in the business environment also forestall innovation. The country ranks at 65 of 190 on Ease of Doing Business, putting it below peers such as Estonia, Latvia, Lithuania, Poland, and Slovenia (EU-5). The number of procedures to start a business is the largest in the comparator group (7 days in Croatia but only 3 in Lithuania and 2 in Slovenia).

The country displays poorer performance when it comes to resolving insolvency than Estonia, Latvia, Poland, and Slovenia but better performance than Lithuania. The recovery rate (cents on the dollar) for Croatia (30.5) is the lowest among the comparator group. Getting credit also appears more challenging in Croatia than in other EU5 economies, with the exception of Slovenia.

Modest Research Excellence and Science—Industry Collaboration

Two other constraints are a drag on innovation: limited research "excellence" and weak science–industry collaboration. Data from the Innovation Union Scoreboard show that the country falls way below the EU average on the share of scientific publications among the top 10 percent most cited worldwide. Croatia had 3.2 scientific publications in 2009, while the EU average was 11. The country is above Latvia, but below the Czech Republic, Poland, the Slovak Republic, and Slovenia (figure 5.27). Further, the average citation impact[13] during 2003–10 was 0.65 for Croatia, but it was 1.31 for the EU. Croatia's main areas of publication in 2012 were medicine; social sciences; agricultural and biological sciences; biochemistry, genetics, and molecular biology; physics and astronomy; and engineering.[14]

Although the number of international scientific copublications per million of population in 2011 was higher than the EU average (405.1 versus 331.3), the country is below the standards of the EU in the share of international publications in total publications. On the first metric, Croatia outperformed countries such as Bulgaria, Latvia, Lithuania, and the Slovak Republic in 2011 but found itself below the Czech Republic, Estonia, and Slovenia. On the second, the share increased from 25 percent in 2003 to about 33 percent in 2012—still far below the EU (44 percent). In addition, the country falls slightly below the EU average on the number of new doctoral graduates per 1,000 population aged 25–34: Croatia has 1.4 new doctorate graduates, against the EU average of

Figure 5.27 Scientific Publications among the Top 10 Percent Most Cited Ones, 2009

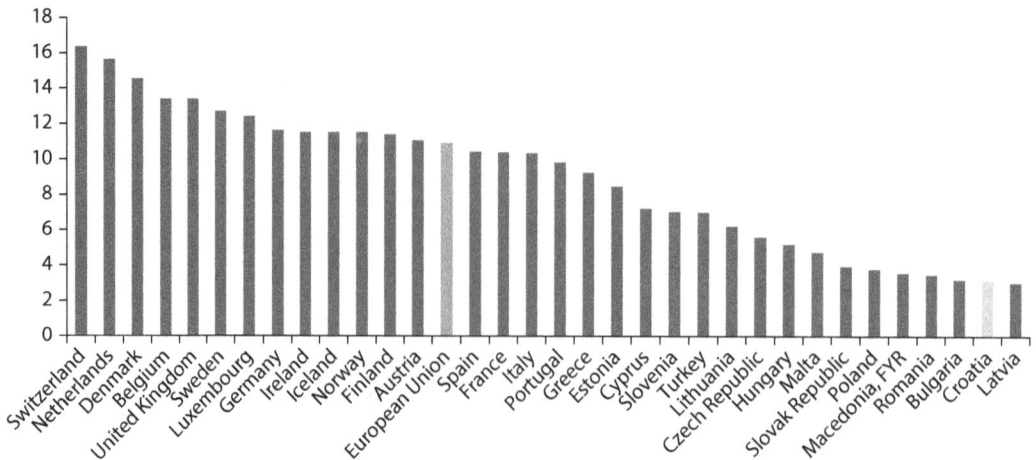

Source: Innovation Union Scoreboard 2014.

1.5 in 2010; it underperforms the Czech Republic, Latvia, and Lithuania, but outperforms Slovenia and the Slovak Republic (figure 5.28). During 2000–11 the number of PhDs increased by 13.8 percent.

Another barrier to innovation is the lack of adequate links between research institutes and the private sector. The country performs poorly on the number of public–private copublications per million of population: 27.4 versus 52.8 for the EU in 2011. Its rate is above those for Estonia, Latvia, and Lithuania, but below those for the Czech Republic and Slovenia (figure 5.29). The share of innovative

Figure 5.28 New Doctorate Graduates (per 1,000 People Aged 25–34), 2010

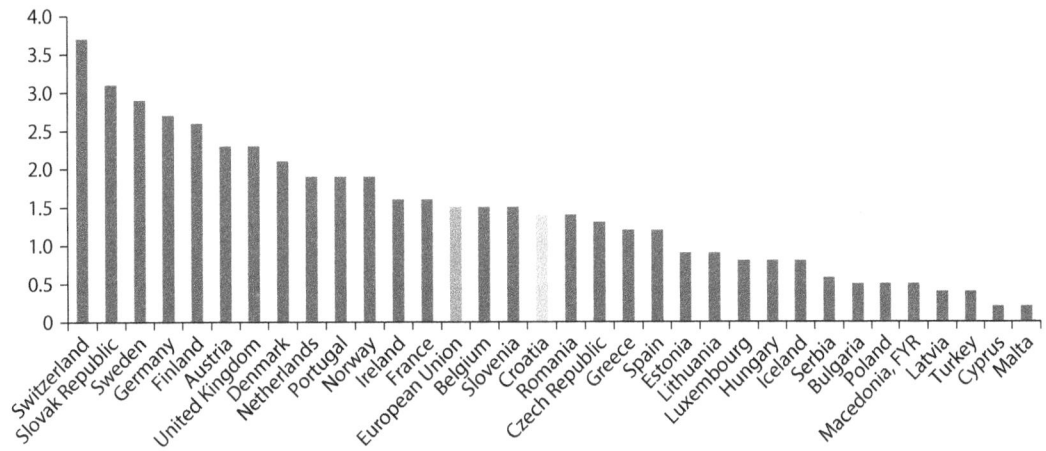

Source: Innovation Union Scoreboard 2014.

Figure 5.29 Public–Private Copublications, 2011
Per million population

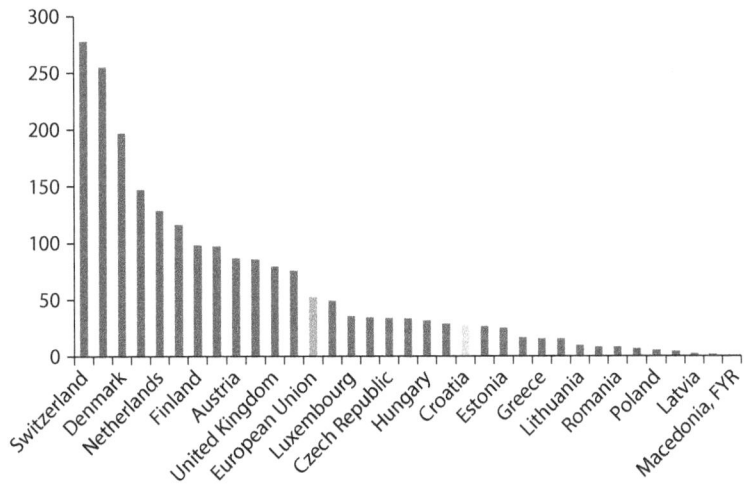

Source: Innovation Union Scoreboard 2014.

Smart Specialization in Croatia • http://dx.doi.org/10.1596/978-1-4648-0458-8

Figure 5.30 Innovative SMEs Collaborating (as % of All SMEs), 2010

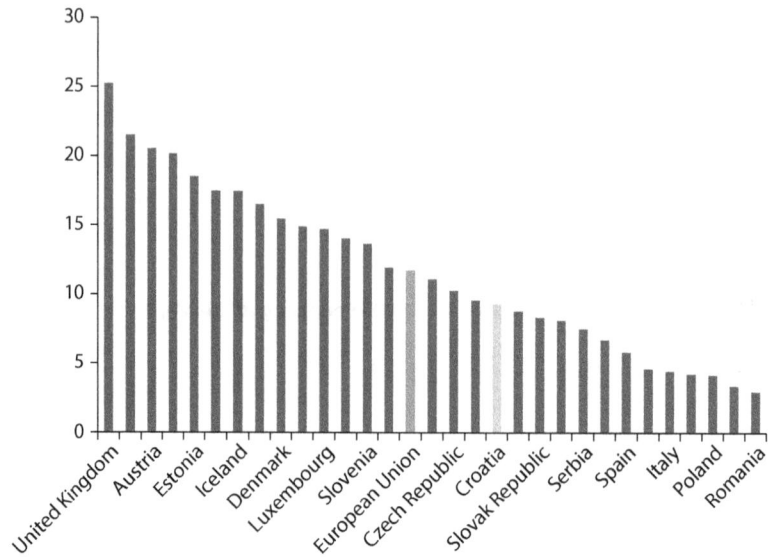

Source: Innovation Union Scoreboard 2014.

SMEs collaborating with others (as a share of total SMEs) is also below the EU average (9.3 percent versus 11.7 percent in 2010) (figure 5.30). Croatia ranks above Poland and the Slovak Republic, but below the Czech Republic, Estonia, and Slovenia.

Cooperative links seem to encourage innovation. According to the Community Innovation Survey (2010), 32.6 percent of Croatian firms that have introduced product or process innovations in 2008–10 conducted innovation activities with partners, while the EU average was 26.5 percent. Cooperation, however, is dominated by large firms (a pattern we saw earlier); SMEs often have weak collaboration networks. According to bibliometric data (SCImago Research Group 2012) database collaborations as measured by copublications between private and higher education institutions in Croatia represent 0.79 percent of total collaborations, or less than half the EU average. A different trend is seen in the uses of scientific information for innovation, however. Universities and R&D companies are rarely seen as sources of information for innovation: only 6.9 percent of firms in Croatia would look to universities for that purpose, and 3.9 percent would seek it from R&D companies, which also shows that Croatian firms are not using domestic knowledge.

Universities in Croatia largely rely on individual initiatives and lack a consistent institutional approach for technology transfer, according to a 2011 report by the European Commission. Most of the universities have neither their own R&D strategy nor technology transfer infrastructure (European Commission TEMPUS 2011). There is still no clear legal or regulatory framework covering intellectual-property and technology-commercialization rights in universities, grounded instead in common law. For example, ownership of such rights for inventions is governed by the Labor Act, which refers primarily to the inventions and relations between inventors (employees) and employers, and protects the rights of employers rather than of employees (WBC-INCO.NET 2011).

Nor are there clear guidelines or a legal framework for spin-offs created by scientists, regardless of whether they are public servants or researchers. Guidance on incentives to researchers to participate in technology transfer activities (e.g., recognition in curricula; researchers' rights to participate in licensing revenues; and equity participation in new firms) is also lacking. Several universities—including the University of Zagreb, University of Rijeka, and University of Split—are developing their own guidelines for intellectual property rights (University of Zagreb 2009).[15]

Among other initiatives in technology transfer, SPREAD (IRCRO) aimed to promote collaboration in the science industry with the provision of matching grants to support joint projects. SPREAD has been designed to foster joint research between private sector and public research organizations with a focus on SMEs. To help establish the collaboration, the program provides matching grants for joint projects between research institutes or universities on the one side, and SMEs on the other. The program stimulates maximum use of infrastructure in scientific research centers and supports industrial companies to substantially increase their R&D activities as well as representing an effective risk-sharing scheme. An assessment by IEZ (2011) showed that SPREAD helped deepen the nature of collaboration between science and industry (box 5.3). Another review of SPREAD by Technopolis (2013) also gives a positive assessment.

Box 5.3 Assessing the Impact of SPREAD (IRCRO)

IEZ conducted an evaluation consisting of an exploratory research at four participants in the program (IEZ 2011), as well as a quantitative research module based on questionnaire answers from eighteen beneficiaries.

As part of the exploratory research, four case studies were developed using information collected through direct interviews. Results showed that:

- Before applying for IRCRO project, companies already had a clear and well-elaborated idea of a new product and were looking for financial opportunities for its development. These projects submitted for IRCRO funding were in line with the company core business and strategic orientation.
- Companies rated the process of submitting applications to IRCRO differently; some had spent more resources in terms of time and personnel involved and considered filling-up of the application forms and reporting to be too time consuming. All the companies reported that having a well-developed product idea before entering the application process was a significant advantage in the application process.

box continues next page

Box 5.3 Assessing the Impact of SPREAD (IRCRO) *(continued)*

- The main motive to apply for IRCRO was affordable financing, followed by the access to R&D services provided by researchers specialized in that particular field.
- Companies welcomed the IRCRO requirement for cofinancing because they felt it added credibility to the company's project, and increased motivation and responsibility. However, matching funds for some companies (micro and newly established) were hard to ensure, especially in the time of recession.

With regard to the quantitative research module, there are 20 projects financed within the IRCRO grant scheme and 18 of them responded to the questionnaire, which represents a high response rate of 90 percent. The main results of the quantitative research show that:

- The two main reasons for firms' participation in the program were to reduce financial risks related to R&D activities and to enhance competitiveness.
- The main reported benefit was access to affordable financing for R&D projects and support for companies to improve their competitive positions both at national and at international levels.
- For about one-third of respondents the grant increased R&D capacity, improved knowledge, and enhanced reputation.
- Without the grant the scope of R&D activities would have been lower and their duration considerably longer. Also, more than half of the companies would not have been able to invest in R&D activities or to implement planned projects.
- Most beneficiaries declared their intention to apply again for the program, which suggests an overall positive response.

Source: Authors' elaboration based on IEZ (2011).

Inadequate Policy Governance

It is critical to address challenges in the coordination of innovation at vertical and horizontal levels in order to achieve consistency of objectives and priorities among multiple strategies and institutions. In most governance systems, the level of information available decreases both horizontally and vertically, particularly in the absence of central coordinating agencies or mechanisms to reduce asymmetric access to information. The results of such an opaque governance structure is a policy-making and implementation system that lacks cohesion and creates misdirected policies and programs that suit the needs of individual agencies or stakeholders and does not adequately improve the overall system. The challenge, therefore, is to set up a structure of incentives that aligns the interest of the "principal" and the "agent" in the process of both policy making and policy implementation.

Perhaps the biggest challenge for boosting research and innovation impact is strengthening policy governance. The system is not at the moment fully functional. Public funds are allocated without clear prioritization or results

Figure 5.31 Government Funding of BERD
Percent of GDP

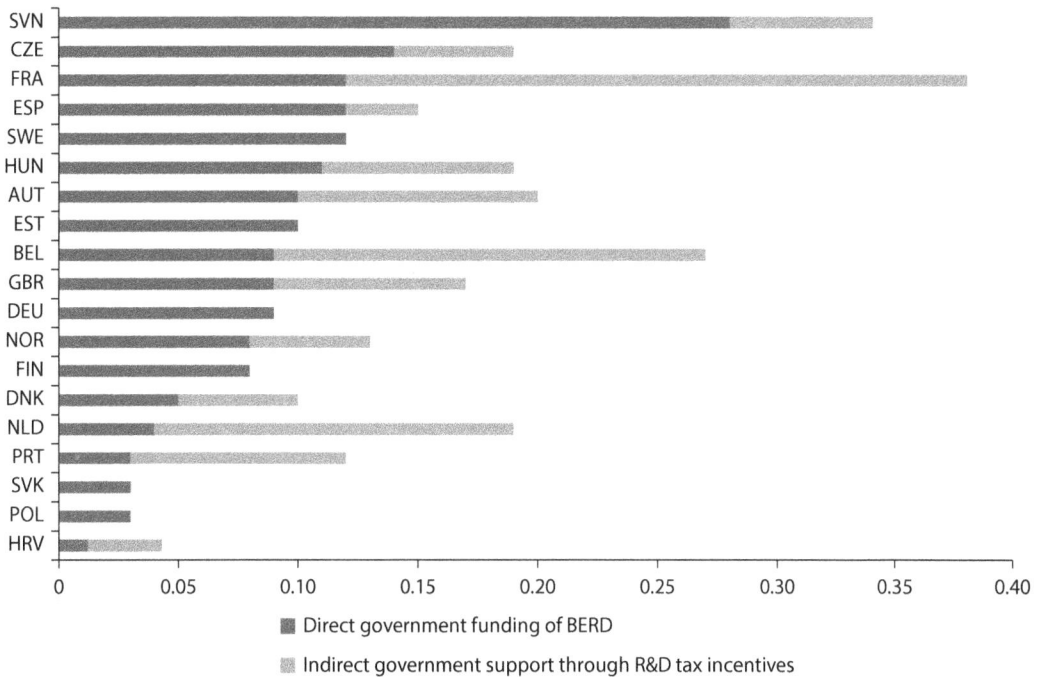

■ Direct government funding of BERD

■ Indirect government support through R&D tax incentives

Source: OECD and Eurostat.

orientation, as reflected in the low government support to businesses and in the low share of experimental research in total research. Croatia has the lowest direct government funding of business R&D within a large set of European countries (figure 5.31). Its indirect government support through tax incentives has more weight, although it still does not exceed 0.05 percent of GDP. Public support to BERD (business R&D) is very small compared with Slovenia (figure 5.32). Further, the share of funds for experimental R&D was quite small in 2012, at under half the U.S. rate (figures 5.33 and 5.34).

Technology and innovation policy is still fragmented in Croatia, ensuring programs with overlapping objectives and nonrationalized resources; and governance challenges may have hindered the efficient management of research institutes after these entities were centralized under the Ministry of Science, Education, and Sports (MSES). Under the Law on Research and Development, MSES is responsible for managing and administering the public research institutes. The law defines three administrative bodies for governing the institutes: the Governing Council, the Academic Council, and the director of an institute. MSES is the administrative body responsible for planning, funding, and monitoring the overall science and education system. The Ministry of Economy, also has programs with a focus on increasing business–industry linkages. These two ministries seem not, however, to coordinate their policies, as both

Figure 5.32 Government Support for BERD
Euro per inhabitant

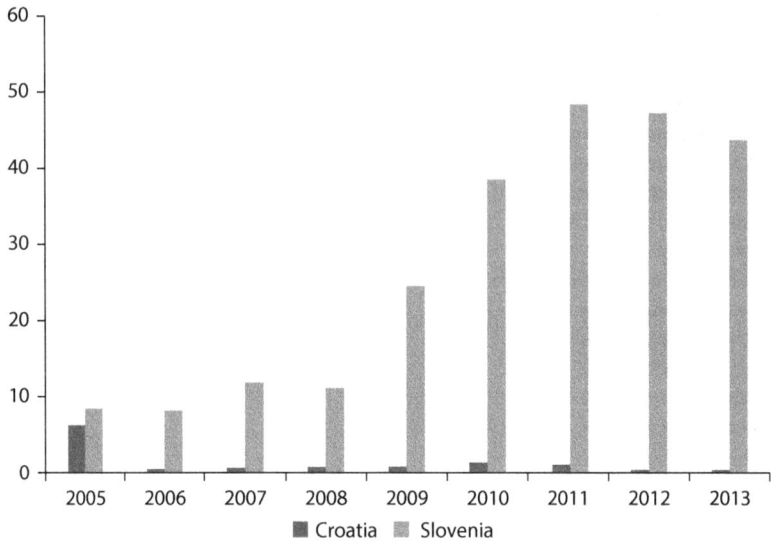

■ Croatia ■ Slovenia

Source: Eurostat.

Figure 5.33 Composition of R&D Expenditures, Croatia, 2012
Percent

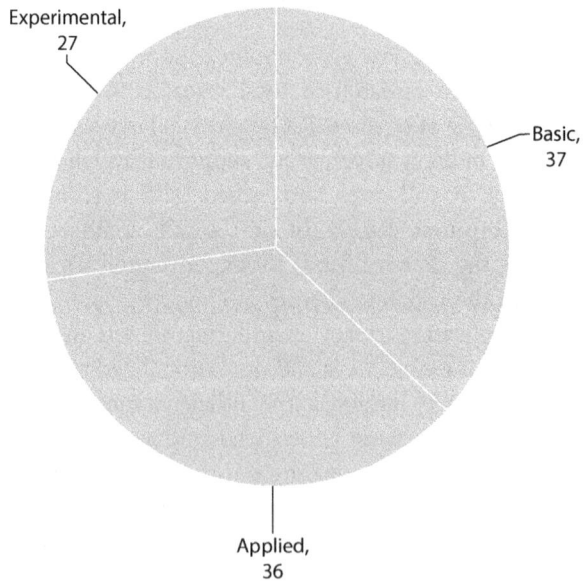

Experimental,
27

Basic,
37

Applied,
36

Source: Eurostat.

Figure 5.34 Composition of R&D Expenditures, United States, 2007
Percent

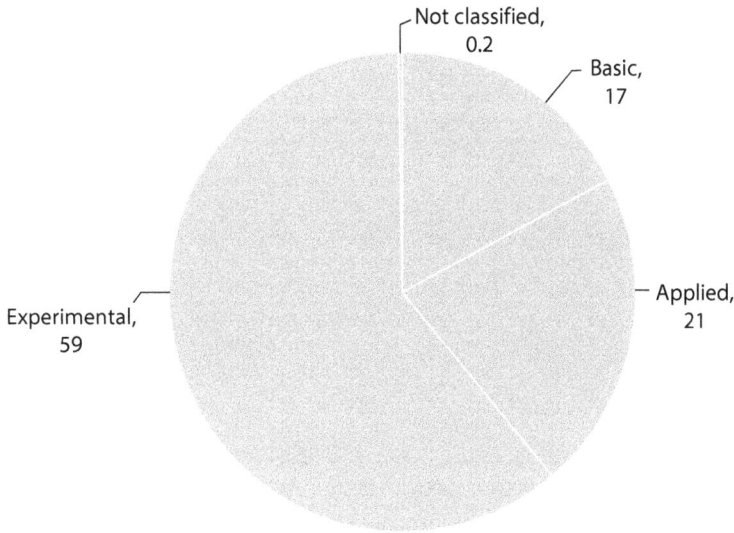

Not classified, 0.2

Basic, 17

Applied, 21

Experimental, 59

Source: Eurostat.

Figure 5.35 Operating Costs and Salaries

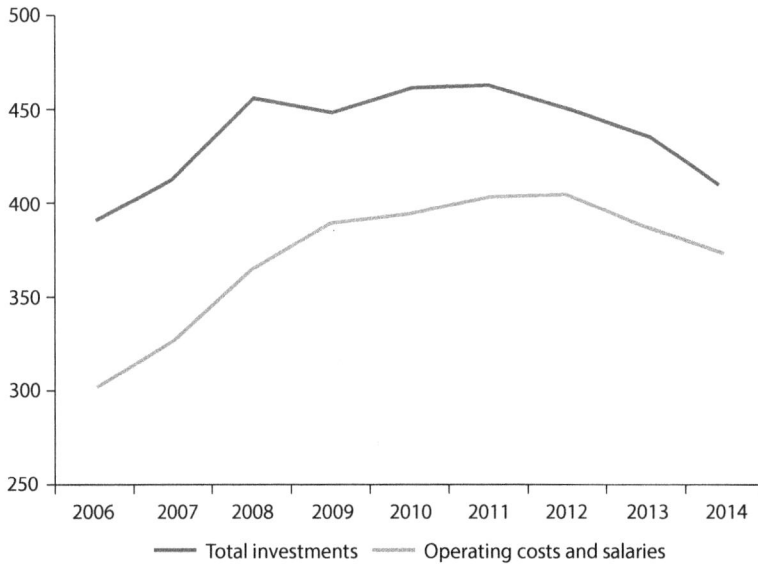

Total investments ▬▬▬ Operating costs and salaries

offer programs for similar objectives and beneficiaries. Moreover, some MSES technology-commercialization programs with very similar objectives are managed by separate agencies.

MSES commands the largest share of public expenditures in R&D, although its R&D spending is mainly used to pay salaries (figure 5.35) and other

Figure 5.36 Capital Investments and Project Financing

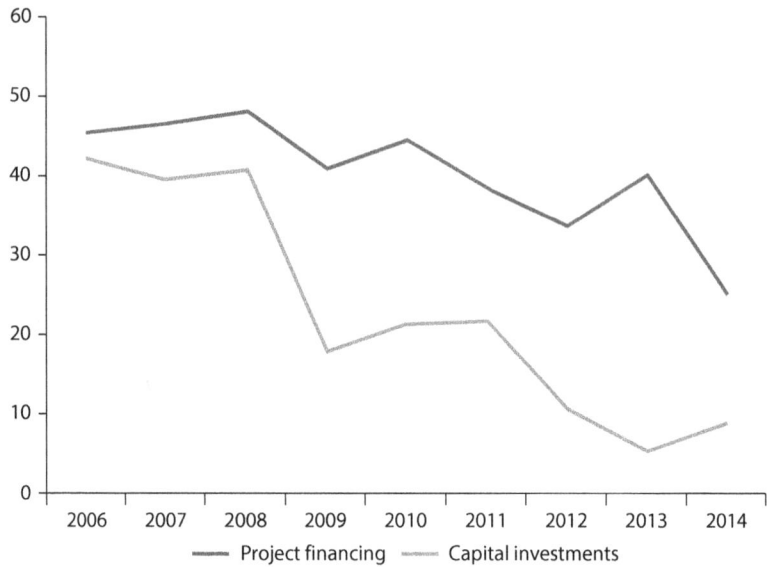

Source: Croatian budget information.

earmarked expenditure. This leaves little for other investment (figure 5.36). Around one-fourth of the overall spending of MSES is allocated on a per-head-count basis to 25 research institutes, which lack, however, the administrative flexibility to manage those resources efficiently, to reward research excellence and productivity, and to sanction recurring underperformance. Having a single organization defining research and innovation programs simply because it is the managerial authority may generate highly inefficient allocation if spending is not coordinated by a single body, such as a national innovation council.

Recent changes may improve matters. In 2013, a new model of dedicated institutional financing for science activity was introduced. Three-year contracts were signed with directors of the public research institutes and the rectors of public universities, enabling financing to be allotted per a transparent set of indicators. In 2013 the Croatian Agency for Science and Education also performed quality evaluation of all research institutes and universities in Croatia with the support of international experts, and the results will be used to improve their financing.

Coordination not just is a national concern, but also cuts across borders. Croatia has, for example, joined the Western Balkans Regional R&D Strategy for Innovation in 2011 with Albania, Bosnia and Herzegovina, Kosovo,[16] FYR Macedonia, Montenegro, and Serbia. The strategy aims to release the innovative potential of the region and address the EU-related opportunities and requirements in research and innovation. The countries will implement programs to improve the research base and conditions for research excellence, promote science–industry collaboration and technology transfer, enable business innovation and innovative start-ups, and strengthen the governance of national research and innovation policies. The strategy recommends a set of programs that build on the recent experience on regional cooperation in research and innovation among Western Balkan countries through bilateral agreements and initiatives funded by the EU. The comprehensive, sectorwide approach adopted as part of this framework complements the programs of other regional initiatives, such as the Danube and Adriatic Ionic Strategies (World Bank 2013).

Good governance implies research systems that are competitive and transparent, with quality-driven recruitment practices and efficient administrative procedures. Better governance of public research institutes and public universities can be achieved through new mechanisms, such as greater use of project funding (typically contracts and grants awarded through competition) and selective increases of funding for research fields linked to socioeconomic needs.

In research institutes, governance mechanisms should encompass performance-driven career development, transparent recruitment policies, and clear rules on ownership and commercialization of intellectual outcomes from research. This also implies that results of publicly funded research are protected and published in a way that encourages their commercialization.[17] Examples of governance principles to help Croatia's policy makers make research careers more attractive can be found in the European Charter for Researchers and the Code of Conduct for the Recruitment of Researchers. Croatia's governance of research institutes needs to ensure research excellence and a balance between institutional funding, which facilitates funding stability, and competitive (project-based) grant funding, which fosters competition in science.

Universities require stronger autonomy to make decisions. Areas include education and training research and innovation, open transparent and merit-driven recruitment methods, institutional accountability, quality assurance, and the ability to access alternative sources of funding and engage in interactions with industry. Competitive education implies more involvement by the business sector in curriculum development and doctoral training, so that skills better match industry needs.

The above issues and steps to resolution show how critical it is to coordinate innovation so as to achieve consistency of objectives and priorities among multiple strategies and institutions. In most governance systems, information is dissipated, horizontally and vertically, with no central coordinating agencies or mechanisms to reduce asymmetric access to it. Such opacity may suit the needs of individual agencies or stakeholders, but it does not marshal resources to benefit the whole system.

Notes

1. Inputs for this chapter come from three World Bank Group reports: "Western Balkans Regional R&D Strategy and Innovation. Country Paper Series: Croatia," by Paulo G. Correa and Pluvia Zuniga (2013); "Ensuring Smart Growth—Innovation

and R&D Policy," by Dragana Pajovic and Paulo G. Correa; "Venture Capital Program for Croatia: A Proposal" (2013) by Paulo G. Correa, Sebastian Penn, Thomas Nastas, and Andrew Myburgh.

2. The EU-8 comprises the Czech Republic, Estonia, Hungary, Latvia, Lithuania, Poland, the Slovak Republic, and Slovenia.

3. Data came from the Community Innovation Survey 2010. This is a survey of innovation activity in enterprises. The harmonized survey is designed to provide information on the innovativeness of sectors by type of enterprises, on the different types of innovation, and on various aspects of the development of an innovation, such as the objectives, sources of information, public funding, and innovation expenditures.

4. We selected Slovenia because the country is one of Croatia's main comparators for the share of business expenditure on R&D.

5. Intensive margin is related to changes in existing activities, whereas extensive margin is related to entry and exit.

6. We employ a different classification following the Enterprise Survey: small firms (0–19 employees), medium firms (20–99 employees), and large firms (100 or more employees).

7. In high-income countries, SMEs are responsible for over 50 percent of GDP and over 60 percent of employment, but in low-income countries, they are responsible for less than half of that: 30 percent of employment and 17 percent of GDP (according to the Harvard Kennedy School, Entrepreneurial Finance Lab Research Initiative). This SME gap is called the "missing middle," a pattern common in developing countries, which often have many microenterprises and some large firms, but far fewer SMEs.

8. European Commission. 2014. "ERAWATCH Country Reports 2013: Croatia."

9. This section is a summary of the report "Venture Capital Program for Croatia: A Proposal" (2013).

10. ERAWATCH. "TEHCRO: Infrastructure for technology transfer" http://erawatch.jrc.ec.europa.eu/erawatch/opencms/information/country_pages/hr/supportmeasure/support_mig_0001

11. Other initiatives include the grant schemes from the National Science Foundation; the MSES Brain Gain—Homing Program; and MSES directly funded projects. See ERAWATCH (2013).

12. From the start, UKF has adopted and used a transparent and meritocratic selection process including anonymous and international peer reviews and a lean governing body of one representative of each of the private sectors, universities, and research institutes. The program required foreign or local research institutions or the private sector to fund at least 10 percent of the total value of the project.

13. Citation counts that are normalized for the field and year of publication.

14. Innovation Union Scoreboard.

15. Efforts are, though, in hand to develop a national policy for creating and managing intellectual property rights at research institutions: the first steps have been taken within a CARDS 2003 project, Intellectual Property Infrastructure for the R&D Sector. Based on this project as well as several TEMPUS projects, such as Fostering Entrepreneurship in Higher Education, offices for technology transfer have been established across Croatia (WBC-INCO.NET 2011).

16. This designation is without prejudice to positions on status and is in line with United Nations Security Council Resolution 1244 and the International Court of Justice's Opinion on the Kosovo Declaration of Independence.

17. Merit-based recruitment implies not only scientific productivity, but also a wider range of evaluation criteria, such as teaching, supervision, teamwork, knowledge transfer, management, and public awareness activities (see Innovation Union and the Code of Conduct for Recruitment of Researchers, European Commission).

Bibliography

Bloom, N., M. Draca, and J. van Reenen. 2011. "Trade-Induced Technical Change? The Impact of Chinese Imports on Innovation, IT and Productivity." NBER Working Paper 16717, National Bureau of Economic Research, Cambridge, MA.

Cohen, Wesley, M., and Daniel A. Levinthal. 1990. "Absorptive Capacity: A New Perspective on Learning and Innovation." *Administrative Science Quarterly* 35: 128–52.

Correa and Hansen. 2010. World Bank.

Correa, P. G., S. Penn, T. Nastas, and A. Myburgh. 2013. *Venture Capital Program for Croatia: A Proposal.* Washington, DC: World Bank.

Correa, P. G., and P. Zuniga. 2013. *Western Balkans Regional R&D Strategy and Innovation. Country Paper Series: Croatia.* Washington, DC: World Bank.

Croatian Private Equity and Venture Capital Association. http://www.cvca.hr/home.

Cusolito, A. 2009. "Competition, Imitation, and Technical Change: Quality vs. Variety." Policy Research Working Paper 4997, World Bank, Washington, DC.

Doing Business 2015 database. http://www.doingbusiness.org/reports/global-reports /doing-business-2015.

European Commission. 2013. "Research and Innovation Performance in EU Member States and Associated Countries." European Commission, Brussels.

———. 2014a. *Innovation Union Scoreboard 2014.* Brussels: European Union.

———. 2014b. *ERAWATCH Country Reports 2013: Croatia.* Luxembourg: European Union.

European Commission. TEMPUS. *Higher Education in Croatia.* http://eacea.ec.europa.eu /tempus/participating_countries/reviews/croatia_review_of_higher_education.pdf.

Eurostat database. http://ec.europa.eu/eurostat/data/database.

IEZ (Institute of Economics, Zagreb). 2011. "Public Sector Research Funding." Innovation Policy Platform, OECD, Paris.

Krstulovic, Hrvoje, and Marko Makek. 2013 "Investment Funds in Croatia: An Overview to Potential Business Partners and Project Intermediaries." http://www.ccbn.hr /repository/investment_forum_presentations/Croatian.

Mohnen, P., and B. Hall. 2013. "Innovation and Productivity: An Update." *Eurasian Economic Review* 3 (1): 47–65.

OECD. 2011. "Changing Patterns of Governance in Higher Education." http://www.oecd .org/education/skills-beyond-school/35747684.pdf.

Pajovic, D., and P. G. Correa. "Ensuring Smart Growth: Innovation and R&D Policy." World Bank, Washington, DC.

Radas, S., Anic, D., L. Bozic, J. Budak, and E. Rajh. 2011. "Evaluation of the Innovation Programs Financed by World Bank in Croatia (Activity A)." Background Paper for the ICR of the Croatia STP. http://www.eizg.hr/en-US/Evaluation-of-the-Innovation -Programs-Financed-by-World-Bank-in-Croatia-662.aspx

Radas, S., and L. Božić. 2009. "The Antecedents of SME Innovativeness in an Emerging Transition Economy." *Technovation* 29: 438–50.

SCImago Research Group 2012 database. http://www.scimagojr.com.

Seker, M. 2011. "Estimating the Impact of R&D and Innovation in Croatia." Background note prepared for the World Bank report "Ensuring Smart Growth—Innovation and R&D Policy."

Švaljek, Sandra. 2012. "R&D Tax Incentives in Croatia: Beneficiaries and Their Benefits." Paper presented at the conference "Hidden Public Spending Present and Future of the Tax Expenses." 117–30. Institut za Javne Financije, Zagreb.

Technopolis. 2013. *Ex Post Evaluation of BICRO's Technology Programmes*. Final Report.

University of Zagreb. 2009. "Rules on Office Transfer Technology." http://technology .unizg.hr/_download/repository/Pravilnik_o_Uredu_za_transfer_tehnologije.pdf.

WBC-INCO.NET. 2011. "Innovation Infrastructures: Croatia." Coordination of Research Policies with the Western Balkan Countries.

World Bank. 2009. *Croatia EU Convergence Report: Reaching and Sustaining Higher Rates of Economic Growth*. Report 48879-HR. Washington, DC: World Bank.

———. 2012. *Implementation Completion and Results Report*. Washington, DC: World Bank.

———. 2013. "Western Balkans Regional R&D Strategy for Innovation: Overview of the Research and Innovation Sector in Western Balkans." World Bank, Washington, DC.

CHAPTER 6

Conclusions and Policy Implications

To boost its economic growth in the coming decades, Croatia needs to raise productivity, deepen trade integration (requiring it to develop its comparative advantage in more skill- and knowledge-intensive sectors), and foster innovation.

Conclusions

Croatia has lost trade competitiveness in recent years and is less integrated in international trade than its peers.

After impressive growth until 2008, per capita GDP has been stagnant since the start of the global crisis. Between 2007–08 and 2011–12, Croatia's export growth was virtually flat; export competitiveness, as measured by export market share change, declined; and the country's export openness lay below the average of peers with similar per capita incomes. This leaden trade performance is an important obstacle for economic recovery and long-term growth.

Lack of structural transformation in the export sector is at the heart of the underwhelming trade performance.

The country faces challenges in increasing its competitiveness in high-quality regional markets and in sustaining growth of more sophisticated products. It lost market share in the traditional EU-15 in the past decade because of severe export competition from low-cost countries, while the overall complexity and sophistication of its goods and services export basket showed only modest gains, such that Croatia is now trailing all its regional peers on this metric.

Croatia is poorly integrated in international value chains. For example, in the transport vehicles, parts, and equipment sector, Croatia missed the foreign direct investment (FDI) rush in the 1990s and early 2000s. This in turn has had large negative effects on technology transfer, as well as on innovation and productivity gains in the productive sectors, as globally GVCs have been an important source of knowledge and incentives for product and process improvements by local providers.

Exports of modern services have not developed enough. Services exports have further concentrated in tourism—its share expanded from 68 percent in 2000 to 73 percent in 2012—and away from modern business services, which are key to increasing productivity in manufactures and to promoting innovation.

Croatia's productive structure has shown just moderate transformation over the past two decades, indicating only limited renewal of the export base, according to product-space analysis. There is specialization in existing markets, while several exports—marginal at present—seem close to current competencies and desirable for their industrial complexity. But the economy has expanded little into new products and markets.

Manufacturing has shown increasing heterogeneity in firm performance but little renewal capacity, contributing to worsening productivity.

Knowledge of firm heterogeneity is particularly important for policy makers to prioritize investments or explore complementarities among firm characteristics. Firm heterogeneity increased over 2008–12 on labor productivity, then capital productivity, and unit labor costs. Across various analytical criteria and controlling for sector-level differences, Adriatic region firms fare worse than Continental firms; trade-integrated firms have better performance than partly or nontrade-integrated companies; small firms perform better than large companies, with higher labor productivity and lower unit labor costs; and SOEs have lower productivity (labor and capital) than private firms—a gap that is widening.

The economy lacks dynamism: the rates of high-growth firms and gazelles (a subset of high-growth enterprises up to five years old) are below Europe and Central Asia (ECA) peers'. These two types of firms are crucial for two reasons: their extraordinary growth can make the largest contribution to net job creation, despite often accounting for only a small share of businesses; and they are a key source of innovation.

Limited entry and exit activity among manufacturing firms is at the heart of Croatia's poor capacity to renew its productive structure.

The lack of vitality of the economy is confirmed when looking at firm entry and exit. The rate of firms new to the market in 2008–12 was lower in Croatia than in the ECA region. Croatia also lags behind on exits: for net entry rates, Croatia presented negative values, indicating that exit outpaced entry. There is some sector variation, however: although all macro sectors have a negative entry rate, it is less so in services, which show strong firm dynamism for both knowledge-intensive services and other less knowledge-intensive services.

Entry and exit activity is in fact reducing aggregate productivity gains.

The economy presented a decrease of 2.88 percentage points in total factor productivity (TFP) over 2008–12. The decomposition of this variation in contributions from survival, entry, and exit firms points to a negative contribution of the net entry effect, which is not what is commonly expected: in principle, the net entry effect should be positive, reflecting the gains arising from the creation of new firms and the exit of obsolete ones. The negative contribution of the net entry effect suggests that the creative destruction process in Croatia is inefficient, as the market might be eliminating firms that are potentially productive (or conversely, preventing the entry of more efficient firms).

The reasons for this negative performance are likely to be linked to anticompetitive regulation and policies in 2008–12 in three areas: still-pervasive state aid in 2011 (even though Croatia had aligned its legislation in the context of EU accession); weak product-market competition (more restrictive than predicted by its development); and specific barriers to firm exit and expansion, notably insolvency resolution and contract enforcement.

Modest innovation performance hinders renewal of the productive structure.

The contribution of innovation to sales growth is 7.2 percent less than the EU-8 average. Labor productivity growth is 9.6 percent lower. The difference is extremely high (32.9 percent) for TFP.

Because non-R&D innovation is acceptable in Croatia, increased business investment in R&D by medium firms is a priority.

Croatia's small and medium enterprises (SMEs) perform much better on innovation in non-R&D activities than in R&D activities, where they are not too far from the EU-27 average for introducing products and process innovations or introducing marketing or organizational innovations. In 2000–11 non-R&D related expenditures grew more than six times to become 8 percent larger than the EU average. But Croatia's performance on BERD (and business enterprise researchers per 1,000 labor force) is much lower than the EU-27 average.

The R&D intensity (R&D spending as a share of turnover) of medium firms is very low in comparison with EU peers, although small and large companies do somewhat better. Another striking feature is the lack of innovation activity among medium firms—the "missing middle."

Factors in the poor performance of business on R&D spending

These factors include concentration of tax incentives by sector and firm size, lack of early-stage financing, poor science–industry collaboration, pervasive weaknesses in the governance framework, and generally underwhelming scores on Ease of Doing Business. The relatively generous tax breaks do not help expand the number of SMEs performing R&D. Large firms receive most of the benefits, although small firms are the majority of beneficiaries. Only a few sectors account for the majority of the tax advantages. And tax breaks are biased toward larger firms. Another drawback is the lack of angel and venture capital services, which risks the premature death of potentially viable innovative start-ups, despite the apparently healthy demand for such funds, while linkages between research institutions and private industry are inadequate, as seen in the low number of copublications. The biggest challenge here is perhaps strengthening policy governance. The system is fragmented, and is not fully functional in that public funds are allocated without clear prioritization and results orientation. The country also needs to have competitive and transparent research systems, with quality-driven recruitment practices and efficient administrative procedures. Finally, the country underperforms on Doing Business indicators such as investor protection, property registration, contract enforcement, and insolvency resolution.

For the EU accession process, Croatia has aligned its legislation on state aid (as well as its antitrust and mergers legislation) with the *acquis communautaire*. However, competition in the domestic market is still weak.

Policy Implications

Croatia's RIS3 presents a good opportunity to push the country out of its low-level equilibrium.

Given the feedback loops seen earlier—poor trade integration discouraging firm-level productivity gains and innovation—Croatia may be trapped in a low-level equilibrium. RIS3 is a good opportunity to spring it from that trap, promoting the conditions for continued structural transformation and the reinvigoration of the productive sector.

As part of the RIS3 process, countries' policy makers have to assess the type of economic specialization of each region and the determinants or binding constraints. The level of available information on the specialization of a region is crucial for the success of targeted policies: ex ante targeting is more likely to achieve its goals if a region's comparative advantage is known to policy makers.

Yet this is not the case for Croatia, partly because it has few clear emerging or existing comparative advantages. This implies that future specializations will be revealed through a flexible strategy encompassing enabling policies for entrepreneurship and market selection, rather than ex ante targeting. The prime goals of Croatia's policy makers may therefore include facilitating firms' access to information, improving market entry and exit conditions, building infrastructure for innovation financing, and helping to build knowledge assets.

In designing and implementing RIS3, policy makers should use a results-based approach in combination with a fully integrated monitoring and evaluating system that allows for structured learning and systematic adjustment of programs and policies toward the predefined objectives. It is essential to replace the emphasis on ex ante definition of sectors and full commitment of resources up front by a results-based approach that allows some flexibility for policy and program experimentation and ex post allocation of resources based on results.

Figure 6.1 presents a suggested framework. It proposes intermediate goals along the way to enhanced productivity and improved innovation (key outcomes), which would lead to a larger and more diversified export base (high-level development objectives), and so to faster GDP and employment growth (final outcome). A summary plan is provided in table 6.1 (discussed in more detail in the subsequent text), followed by an outline of the sector case studies.

Figure 6.1 Framework for the Design and the Monitoring and Evaluation of RIS3 in Croatia

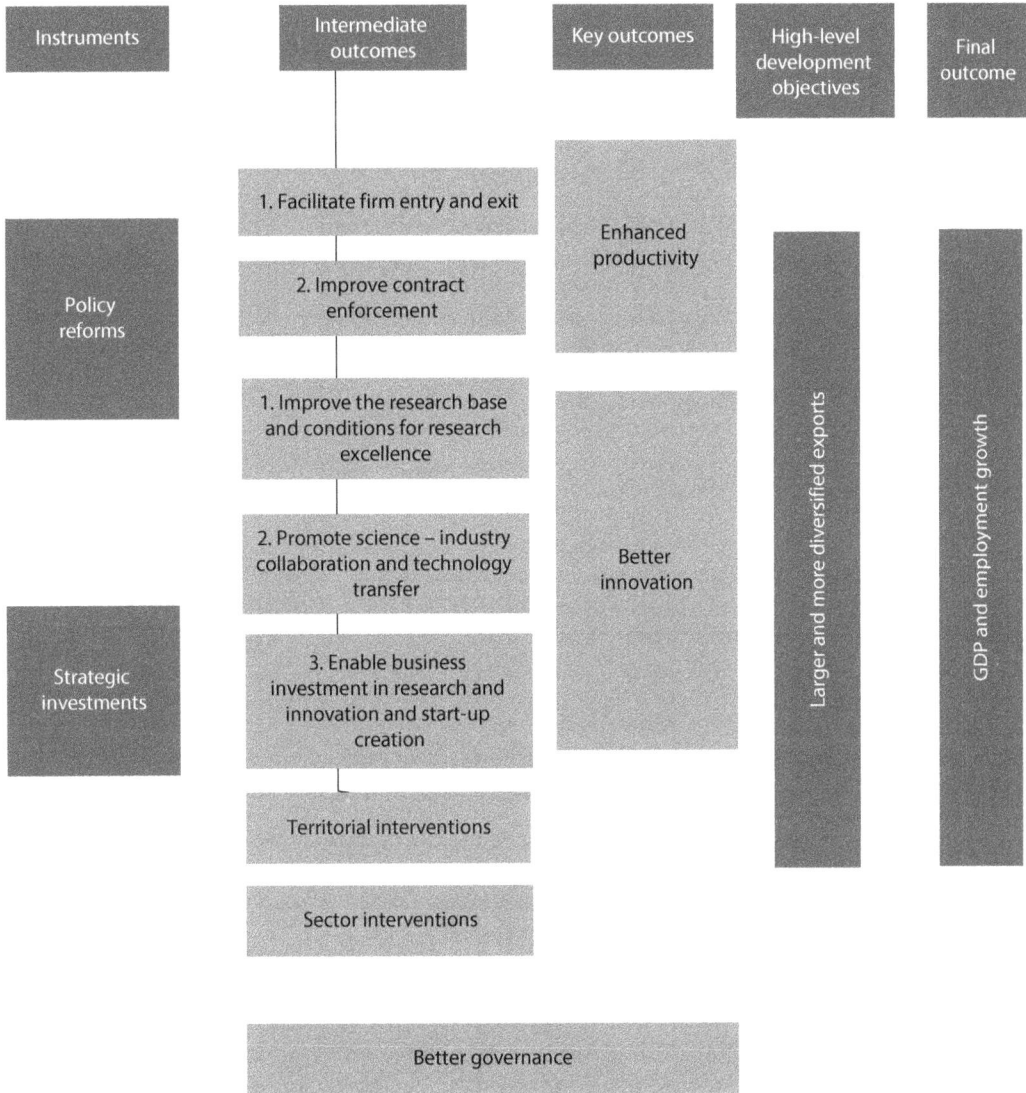

Instruments	Intermediate outcomes	Key outcomes	High-level development objectives	Final outcome

	1. Facilitate firm entry and exit	Enhanced productivity		
Policy reforms	2. Improve contract enforcement			
	1. Improve the research base and conditions for research excellence		Larger and more diversified exports	GDP and employment growth
	2. Promote science – industry collaboration and technology transfer	Better innovation		
Strategic investments	3. Enable business investment in research and innovation and start-up creation			
	Territorial interventions			
	Sector interventions			

Better governance

Addressing weaknesses in the business environment and promoting entry and exit reforms to enhance productivity

Croatia should strive to reduce sector-specific use of state aid to minimize distortions in the economy and facilitate competition. Given the key role of services as intermediate inputs for many firms, the authorities should ensure full implementation of the Services Directive of the European Commission and remove remaining barriers to FDI in services. It would also help to increase firm productivity by improving insolvency procedures through reducing the time for

Table 6.1 Summary Matrix of Potential Areas for Policy Actions and Strategic Investments

Intermediate goals	Specific areas for policy actions and strategic investments
Facilitate firm entry and exit	Reduce sector-specific use of state aid to minimize distortions and facilitate competition
	Complete the implementation of the Services Directive of the European Commission and remove existing barriers on FDI in services
	Improve insolvency procedures: reduce the time for debt recovery and increase the recovery rate
Improve contract enforcement	Reduce the time required to enforce a contract
Improve the research base and conditions for research excellence	Consolidate (restructure) the public research organizations to reduce overhead costs and rationalize the use of public resources
	Introduce performance-based contracts for public research organizations, including defining "mission-driven" research and creating better conditions for efficient management of public research organizations
	Strengthen the connection with the global scientific community, including expanding UKF and improving conditions for researchers' mobility
	Advance reforms in regulating the research profession to further emphasize excellence (e.g., emphasizing publication in high-impact journals and creating better conditions for talented young researchers)
	Introduce measures to increase the retention of skilled labor
Promote science–industry collaboration and technology	Set up the regulatory base for technology transfer infrastructure
	Streamline the regulatory framework for intellectual property rights and technology transfer
	Set up incentives for researchers to participate in technology transfer activities: recognition in curricula, rights to participate in licensing revenues, and equity participation in new firms
Enable business investment in research and innovation and start-up creation	Simplify procedures for firm entry (see above)
	Introduce guidelines for spin-offs from universities
Sector/territorial interventions	Support producers' efforts to obtain international certification (categorization, minimum quality standards)
	Support the creation of associations of producers
	Support branding and marketing
	Support research commercialization
Improved governance	Emphasize use of transparent and competitive allocation of research funds (including international peer review, objective selection criteria, and timely calls for proposals)
	Stress use of transparent and competitive selection of infrastructure research projects (including international peer review, objective selection criteria, and timely calls for proposals) to be implemented through European Structural and Investment Funds
	Continue improving management of research infrastructure, including the adoption of the Infrastructure Road Map (EU mandated) and a policy of open and effective access to those resources by the scientific community
	Strengthen the functioning of policy coordination bodies at the level of prime minister (e.g., National Research and Innovation Council), with participation of the private sector, the creation of a Technical Secretariat, and the adoption of a results-driven monitoring and evaluation framework for public expenditures on R&D and innovation
	Strengthen HAMAG-BICRO as the agency in charge of promoting R&D investments in SMEs and corresponding programs
	Continue to support the implementation of the Western Balkans Regional R&D Strategy for Innovation

debt recovery and increasing the recovery rate, reducing the time needed to enforce a contract and improving contract enforcement.

Improving innovation by supporting R&D investments in SMEs and better performance of the research system

Croatia needs to increase, gradually but substantially, direct support to business investments in R&D. This can be done by HAMAG-BICRO's Proof of Concept and RAZUM programs. In addition, the authorities should consider developing a new program, which could be financed by European Structural and Investment Funds, to support angel and venture capital services. The focus of these funds in innovation should be to promote business expenditures on R&D by young firms and SMEs.

A simple extrapolation exercise shows how much the government would have to increase budget expenditure for direct support of research and innovation activities (table 6.2). Assuming a best-case scenario elasticity of private R&D spending to government spending of about 2, and taking into account GDP levels and the volume of European Structural and Investment Funds for the upcoming programming period, direct government support for private R&D should aim to reach at least 0.21 percent of GDP by 2020.

Table 6.2 Estimated Additional Expenditure Needed to Meet the EU 2020 Target on R&D Spending

EU financing provided through the operational program (including R&D)		
Priority axis 1: Strengthening the economy through application of research and innovation	€534,792,165	
Priority axis 2: Use of information and communication technologies	€269,441,037	
Priority axis 3: Business competitiveness	€997,457,755	
Total (2014–20)	€1,801,690,957	
Yearly estimated allocation (2014–20)	€257,384,422	
Yearly estimate estimated allocation (% of GDP)	0.59%	
Gross domestic expenditure on R&D (% of GDP)	*Current*[a]	*Target*
	2012	2020
	0.75	1.40
Additional % of GDP needed to reach the target	0.65	
Elasticity of direct government support for private sector R&D	2	Based on the Sayek (2009) model[b]
Additional financing needed to reach the EU 2020 target	*0.21% of GDP (government) + 0.42% of GDP (private sector) = 0.65% of GDP*	
0.21% of GDP (direct support for private R&D)	36% of EU financing for competitiveness or €92 million (approximate additional yearly allocation)	

a. Based on available Eurostat data.
b. "Social returns to R&D in Turkey," Selin Sayek Böke 2009.

A degree of support of the enterprise sector is foreseen in the Croatia 2014–2020 Operational Program for Competitiveness and Cohesion, including support for the acquisition of new manufacturing technologies, equipment, and machinery; for technology transfer processes from scientific and research organizations; and for greater use of key enabling technologies for developing new products, services, and business models.

Croatia needs to improve research excellence. Several policy initiatives can be considered to streamline the innovation framework, including the following: strengthening the connection with the global scientific community (for example, expanding UKF and improving conditions for mobility of researchers); advancing reforms in regulating the research profession to emphasize research excellence (such as publication in high-impact journals) in career development; and promoting better access to research infrastructure through an open-access policy.

Several strategic investments designed to enhance the country's innovation system are planned as part of the Operational Program. These include investments in R&D infrastructure and equipment in public and private research institutions and organizations; investments in furnishing, construction, and initial operating costs of private and public institutions supporting commercialization of innovation and technology transfer; establishment of highly focused centers of competencies; establishment of a high-technology network for industry investments to ensure access to scientific databases, scientific publications and journals, and digital resources; and development of a system for professional services for knowledge and technology transfer, awareness raising, and brokerage activities related to the benefits of intellectual property protection and technology transfer in public academic and research institutions.

It is essential to improve links between research institutions and the private sector. Key policy actions include the following: setting up the regulatory base for technology transfer infrastructure; streamlining the regulatory framework for intellectual property rights and technology transfer; and setting up incentives for researchers to participate in technology transfer activities, such as recognition in career development, rights to participate in licensing revenues, and equity participation in new firms.

The Operational Program also encompasses investments that can help promote collaboration between scientific institutions and industry. Croatia plans to support centers of excellence, which should address industry demands; support development of collaborative projects within university–industry initiatives led by demand from industry; and support joint research initiatives for all types of innovation.

The government may also wish to consider the following:

• Gradually increasing funds to HAMAG-BICRO's programs to promote collaborative projects within university–industry initiatives led by demand from industry.
• Supporting joint science–industry collaboration through the promotion of centers of competence (selected on a competitive basis).

Strengthening Governance

Strengthening policy governance seems to be the biggest challenge for boosting research and innovation impact, both on the overall management of European Structural and Investment Funds and on implementation of the policy reforms required to support programs financed by these funds.

The governance framework for research and innovation policy needs greater efficiency to increase the spending impact. The government could consider the following:

- Strengthening the functioning of the policy-making and coordination body at the level of prime minister (e.g., National Research and Innovation Council) with participation of the private sector.
- Creating a Technical Secretariat and adopting a results-driven monitoring and evaluation framework for public expenditures on research and innovation.
- Strengthening HAMAG-BICRO as the agency in charge of promoting R&D investments in SMEs and corresponding programs.
- Accelerating the adoption of the Infrastructure Road Map, which is an EU requirement, and of a policy of open and effective access to those resources by the scientific community.
- Improving merit-driven selection processes, subject to international peer review and transparency.
- Promoting proper impact assessment, public consultations, and the systematic review of policies and programs.
- Speeding up implementation of the Western Balkans Regional R&D Strategy for Innovation.

The governance of research institutions also needs to be improved, including the following:

- Consolidating (restructuring) public research organizations to reduce overheads and rationalize use of public resources.
- Introducing performance-based contracts for public research organizations, including defining mission-driven research.
- Increasing the autonomy and accountability of research organizations.
- Achieving a better balance between institutional funding, which facilitates funding stability, and competitive (project-based) grant funding, which fosters quality competition in science. This would imply an increase in competitive project-based funding and in institutional funding to strengthen the use of performance-based contracts.[1]

The findings of this report have implications for preparing Croatia's RIS3, such as the desirability of the following:

- Replacing the emphasis on ex ante definition of sectors and full commitment of resources up front by a results-based approach that allows some flexibility for policy or program experimentation and ex post resource allocation based on results.

- Adopting gradualism when expanding programs.
- Adopting a fully integrated monitoring and evaluating mechanism for the design and implementation of the research and innovation strategy, allowing for structured learning and systematic adjustment of programs and policies toward the predefined objectives.
- Considering the proposed actions in the context of the National Reform Program—the first such program that the government, as a member state of the EU, adopted (on April 24, 2014).

Note

1. OECD Policy Mix Database: http://stats.oecd.org/Index.aspx?DataSetCode=IPM _STIO.

Examples of Potential Areas for Research and Innovation Specialization

Appendix Summary

The process of developing RIS3 needs to be interactive, regionally driven, and consensus-based. This is because most innovation does not occur in isolation but depends on critical inputs from a wide range of actors. The capacity to absorb, generate, and exchange knowledge at country, regional, and firm levels needs to be developed fast and efficiently to maintain knowledge relevance and to accelerate economic growth.

Taking into account local stakeholders' knowledge and empowering key local actors to shape an intervention's priorities are essential for program success. Each local actor has in-depth knowledge of the strengths and weaknesses of the regional economy, and these views should be subjected to critical review through a combination of analysis and broad-based consultation. Allowing all participants the opportunity to shape local priorities is fundamental to ensuring their commitment to policy implementation (European Commission 2012).

The analysis in this appendix is a starting point for developing pilot RIS3 projects locally. The assessment of four potential areas can be the basis for local economic development strategies, used to address priority issues, capitalize on limited resources, and provide insights for policy makers—if the strategies are properly monitored and evaluated.

The previous chapters have shown that Croatia has a key problem—the lack of renewal and transformation of the manufacturing base—which partly explains the country's stagnant export performance and its low levels of aggregate productivity in some sectors. Illustrating the implementation of RIS3 in a context with latent comparative advantages or scarce information, this appendix presents four case studies that show the challenges and opportunities of fostering regional activities with growth potential. Drawing on desk research, policy dialogue, and focus group discussions, the studies discuss clean energy production, oysters, Slavonski kulen (a traditional meat product), and biotechnology and pharmaceuticals.

One of the most important challenges in the design of RIS3 is selecting the sectors on which the strategy should rely, and the region, of which there are three types:

- Regions with RCA, where targeted research and innovation policies will complement existing productive assets, helping firms to maintain a competitive edge in the sector by investing in R&D or to regain competitive advantage lost to new players in the global market.
- Regions with latent comparative advantage, where the required knowledge partially exists typically, but not only, due to availability of a nontradable, location-specific input, such as a natural resource, or an immovable asset (land and climate, for example), or local common knowledge about the economic activity, including traditions in the region that indicate potential for specialization.
- Regions with unclear specialization, where there is no obvious local asset that induces an economic specialization (see figure 2.2).

Other criteria should also be considered to select the sector, including the mechanism for value creation, endowment of immovable assets, availability of resources, and information about demand and supply conditions.

The Clean Energy Sector: Large Potential for Croatia[1]

The Clean Energy Market

The concept of clean energy covers several areas: renewable energy, carbon capture and storage, energy efficiency, biofuels, electric vehicles, efficient coal technologies, nuclear power, and fuel economy (table A.1).[2]

Clean energy offers multiple benefits. It is crucial to reducing global carbon emissions; contributes to enhancing energy security and resource efficiency; helps reduce local environmental impacts; provides further options and solutions for energy access; and promotes green economic growth. It is needed more than ever: the carbon intensity of global energy supply has hardly improved in 40 years (figure A.1).

Clean technology is a huge global market. In 2011–12 the sector was a US$5.5 trillion global market, and is forecast to grow at around 4.1 percent annually until 2015/16, or significantly faster than the global economy. In 2012, clean tech investment rose by 19 percent in developing countries. By 2012 Africa had the highest growth rate, 6.5 percent, followed by Europe with 3.9 percent, and Asia with 3.7 percent (figure A.2). Despite the recent contraction, global investment in clean energy climbed steeply after 2004 (figure A.3). The trend and volume of new investments varied by region (map A.1): they fluctuated widely in the United States, but rose steadily in Europe and Asia and Oceania. Investments are still very small in Central and South America, as well as in the Middle East and Africa.

Table A.1 Typology of Clean Energy

Clean Energy Areas	Technology/Product	Examples of subcomponents/products
Renewable Energy	Power generation technologies:	Solar photovoltaic technology: panel (e.g.; thin film solar cells), tracking system, inverter, surge protectors Wind generation: turbines, blades, tower Hydropower generation: turbines Biomass-based electricity and heat
	Biomass fuel production	Agricultural residues, waste or wood pellets
	Renewable energy resource mapping	ICT specialized in software/data mining for geographic cadasters
	Lighting	Efficient lamps: CFLs, LEDs, Organic LEDs Programmable LEDs with systemwide monitoring capabilities
	Heating, Cooling, Ventilation	A wide spectrum of emerging smart technologies in HVAC
Energy Efficiency	Real time measurement/ monitoring	Intelligent/smart sensors/high grade sensors/wireless sensors ICT: Web-based monitoring tools, enhanced data analytics
	Appliances	Smart appliances
	Materials	Insulation and construction materials
Smart Grids	Technological components	Advanced Metering Infrastructure Smart/electronic meters Intelligent/Smart transformers Intelligent/Smart inverters Intelligent/Smart sensors (collect data on temperature, vibration, or other)
	Programs/Software	Software: wide diversity of programs to increase efficiency of overall power system (system optimization, demand response, dynamic pricing, forecasting)
	ICT	Information-based controllers (collect data from smart sensors) Information assemblers: displayers and assessors Real-time measurement and monitoring
Electric Vehicles(EV)	Vehicle	Electric engine Mechanical engineering design Electronics Power train
	Vehicle battery	EV Battery System
Electricity Storage	Batteries	Lithium-ion batteries Lead acid batteries Other: flow batteries, thermal storage, flywheels, super-capacitors, sodium/sulphur batteries
CCS	Various operation units	Equipment for capturing (absorbers), separation (distillation systems), compression, pipelines, stacks, etc.

R&D Performance in Clean Energy

Croatia has an ambitious energy strategy. This has three objectives: to secure energy supply; to promote a competitive energy system; and to develop a sustainable energy sector. The country also has treaties with the EU on renewable energy (Directive 2009/28/EC) and environment-energy efficiency. The National Renewable Energy Action Plan has two tough targets for 2020: 20 percent of renewable energy in total final energy consumption; and 35 percent of renewable energy in power generation. There is also a National Energy Efficiency Action

Smart Specialization in Croatia • http://dx.doi.org/10.1596/978-1-4648-0458-8

Figure A.1 Carbon Intensity

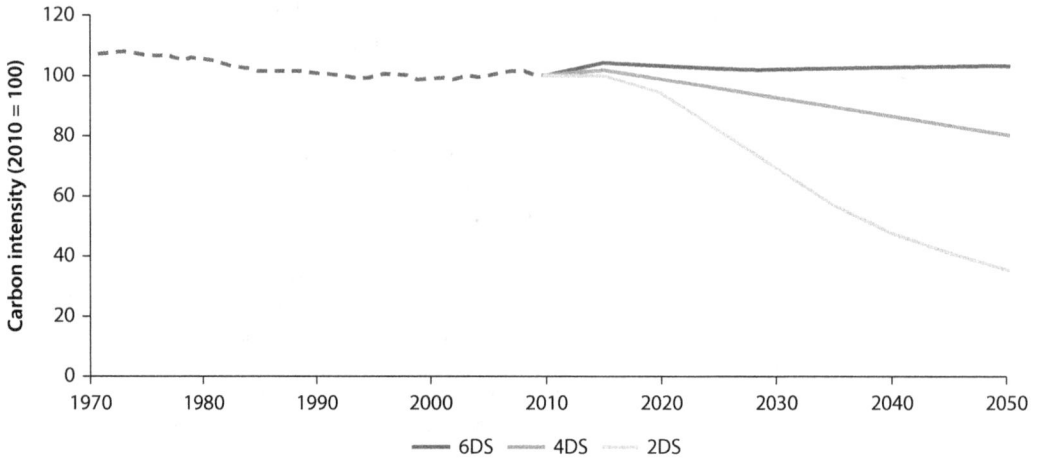

Sources: IEA 2013.
Note: The ETP scenarios (2DS, 4DS, and 6DS) are defined in box 1.2 figures and data that appear in this report can be downloaded from http://www.iea.org/etp/tracking.

Figure A.2 Growth in Clean Tech Sales by Region, 2012

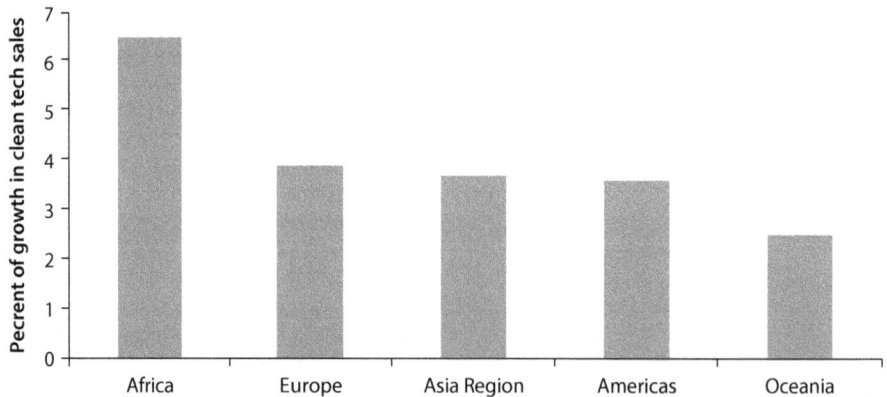

Plan, which promotes energy consumption savings in the residential, tertiary, industry, and transport sectors, as well as the second commitment period of the Kyoto Protocol, which aims to see greenhouses gases cut by 20 percent compared with 1990.

Despite the importance of clean energy, Croatia shows lackluster R&D performance. The number of its clean energy patents has shot up in the last two decades (figure A.4), especially since 2011, but the country is far behind many other countries in the region (figure A.5). Patents in renewable energy generation account for the largest share, followed by those in fuel production from biomass and waste.

Figure A.3 Global Investment in Clean Energy
US$ billion

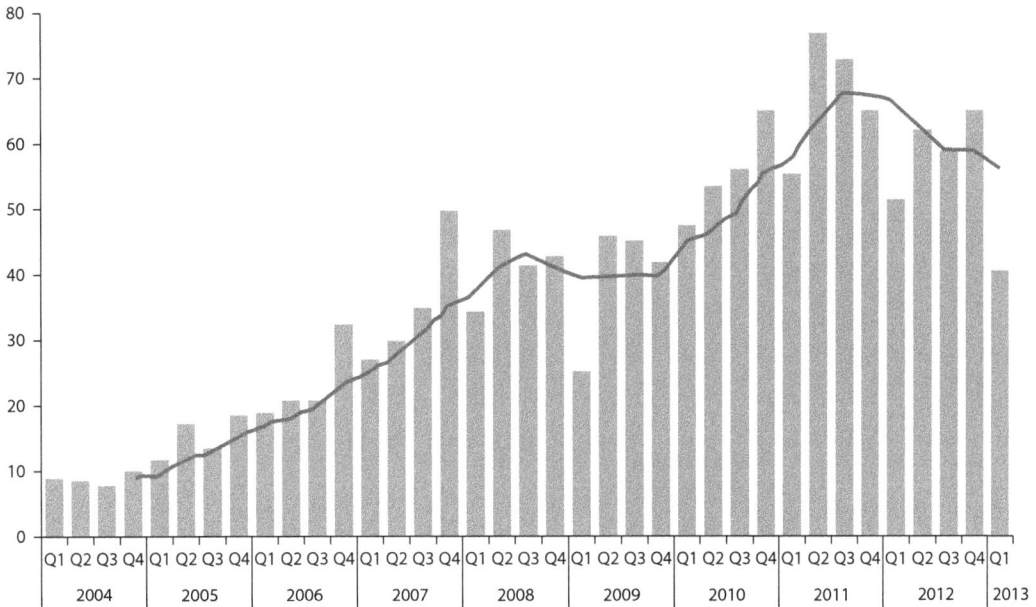

Sources: World Bank 2014; Bloomberg New Energy Finance 2013.

R&D in engineering and technology as a share of R&D spending is growing at only a moderate pace, and the share it represents in total R&D is below that in, for example, the Czech Republic, Hungary, Slovenia, and Poland (figure A.6). Enrollment in engineering, manufacturing, and construction is relatively low, though some signs of growth are emerging (figure A.7).

A New Niche
High-tech exports in areas relevant to clean energy declined in Croatia after the global financial crisis but seemed to start recovering in 2012. Croatia's exports in these areas have been large in electronics and telecommunication products, but have shrunk in electrical machinery and computers (figure A.8). Also, relative to regional peers, they have been modest in value (figure A.9).

Croatia appears to be increasing its exports in power-generating machinery and equipment as well as in telecommunications technology. These areas are extremely pertinent to renewable energy, energy efficiency, and smart-grid technology. Exports in these areas as a share of total exports increased by 6.5 percent between 2011 and 2012 (figure A.10).

Croatia's export basket composition and performance show that industrial machinery and electronic products—closely associated with clean energy technology—are one of the five sectors in which the country specializes. The economy is having emerging success in several products in the sector, including

Map A.1 New Investments in Clean Energy by Region

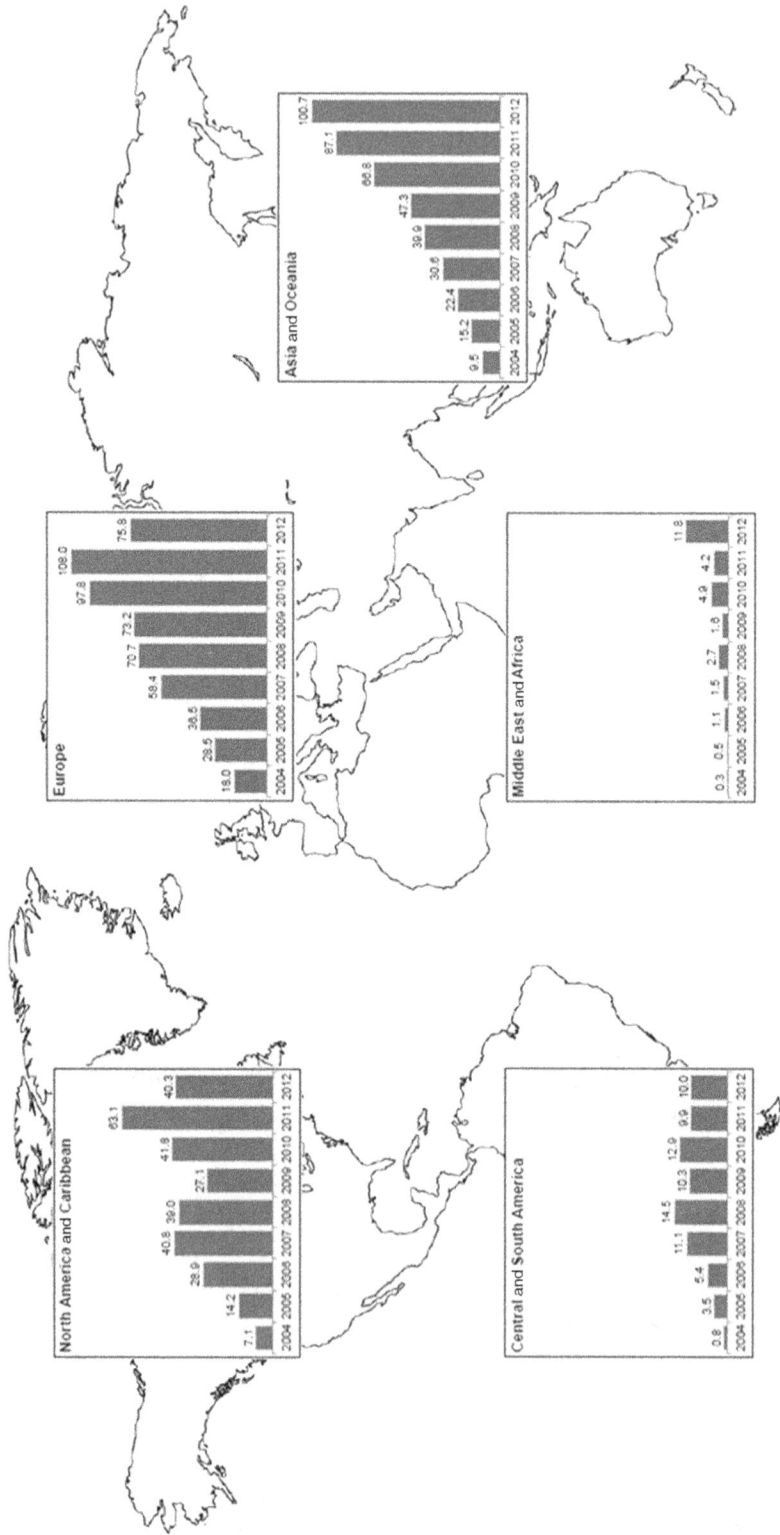

North America and Caribbean

2004	2005	2006	2007	2008	2009	2010	2011	2012
7.1	14.2	28.9	40.8	39.0	27.1	41.8	63.1	40.3

Europe

2004	2005	2006	2007	2008	2009	2010	2011	2012
19.0	28.5	36.5	58.4	70.7	73.2	97.8	108.0	75.8

Asia and Oceania

2004	2005	2006	2007	2008	2009	2010	2011	2012
9.5	15.2	22.4	30.6	39.0	47.3	66.8	87.1	100.7

Central and South America

2004	2005	2006	2007	2008	2009	2010	2011	2012
0.8	3.5	5.4	11.1	14.5	10.3	12.9	9.9	10.0

Middle East and Africa

2004	2005	2006	2007	2008	2009	2010	2011	2012
0.3	0.5	1.1	1.5	2.7	1.8	4.9	4.2	11.8

Source: Bloomberg New Energy Finance 2013.

Figure A.4 Number of Patents in Clean Energy, Croatia

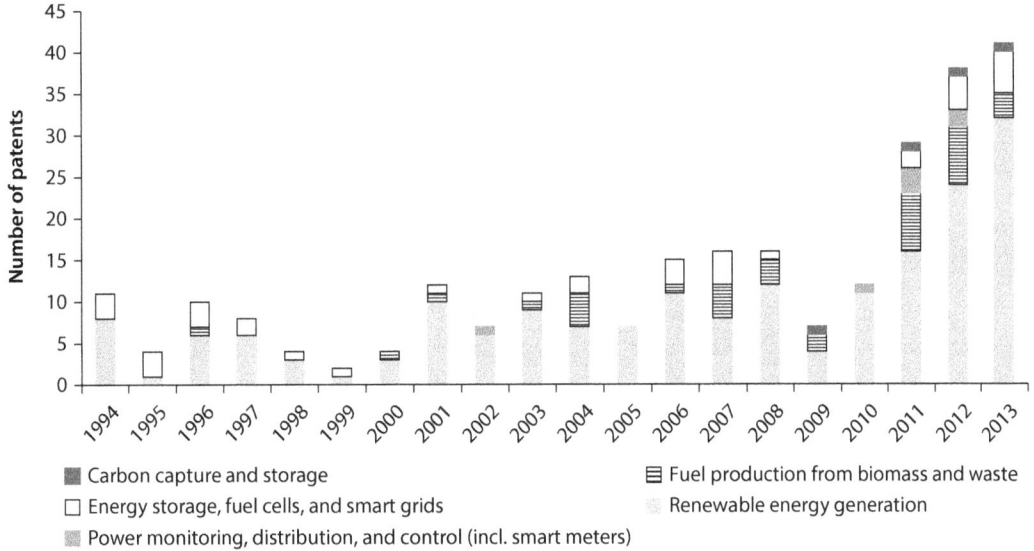

Legend:
- Carbon capture and storage
- Fuel production from biomass and waste
- Energy storage, fuel cells, and smart grids
- Renewable energy generation
- Power monitoring, distribution, and control (incl. smart meters)

Source: Elaborations based on European Patent Office database.

transformers, electric wire, parts of electric power machinery, switchboards (including relays and fuses), and specialized industry machinery (and see chapter 3, specifically table 3.5).[3] Most of these products exhibit a large export share and high complexity.

About 34 clean energy–related exports have a significant or emerging comparative advantage. Identifying high-value exports begins with a stock-taking of the relative performance of export products, measuring performance of sustained exports over 2002–12; and of products with RCA greater than 1. An analysis of HS codes at the 6-digit level in a sample of products (including all renewable energy codes and a basket of energy efficiency and smart-grid components) shows at least 34 export products with significant or emerging RCA (table A.2).

Figure A.11 shows the product-space map with classic smart products in clean energy technology that have RCA higher than 1 in both 2001–02 and 2011–12 at the SITC 4-digit level. These products are located in desirable clusters of the product space where exports share similar productive knowledge.

Roughly 117 marginal exports are not as competitive (indicated by RCA less than 1 in both 2001–02 and 2011–12), but still earn more (export value) than many emerging and significant exports. Several currently marginal exports in clean energy are close to existing competencies and desirable for their industrial complexity. They include electric meters, wind power electric generating sets, photovoltaic cells, silicon material (used in the manufacture of photovoltaic technology), electrical glass insulators, and light-emitting diodes. The sector offers much potential, and efforts should be focused on expanding the production of goods in the sector that are close to the export basket.

Figure A.5 Number of Patents in Clean Energy, Regional Countries

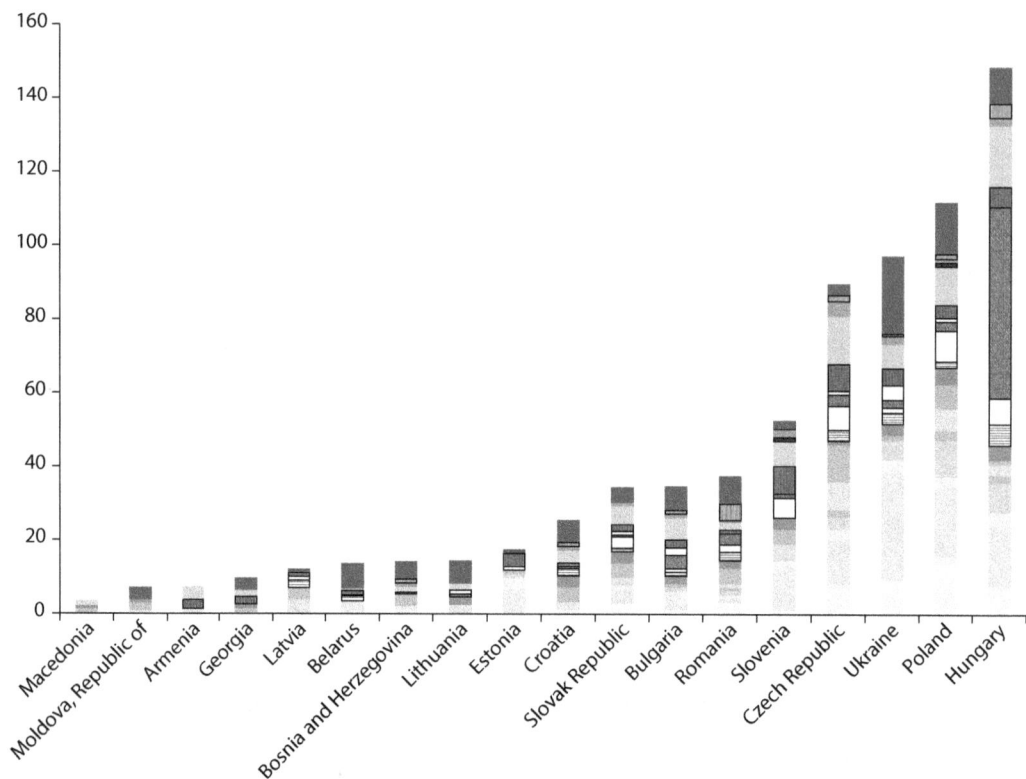

Eastern Europe

- Wind energy
- Technologies specific to propulsion using electric motor (e.g., electric vehicle, hybrid vehicle)
- Technologies specific to hybrid propulsion (e.g., hybrid vehicle propelled by electric motor and internal combustion engine)
- Solar thermal-PV hybrids
- Solar thermal energy
- Solar photovoltaic (PV) energy
- Marine energy (excluding tidal)
- Lighting (incl. CFL, LED)
- Insulation (incl. thermal insulation, double-glazing)
- Hydrogen production (from noncarbon sources), distribution, and storage

- Hydro energy-tidal, stream or damless
- Hydro energy-conventional
- Heating (incl. water and space heating; air-conditioning)
- Geothermal energy
- Fuel from waste (e.g., methane)
- Fuel cells
- Energy storage
- CO_2 capture and storage (CCS)
- Capture and disposal of greenhouse gases other than carbon dioxide (incl. N_2O, CH_4, PFC, HFC, SF_6)
- Biofuels

Source: Elaborations based on OECD Patent Database.

Figure A.6 R&D in Engineering and Technology

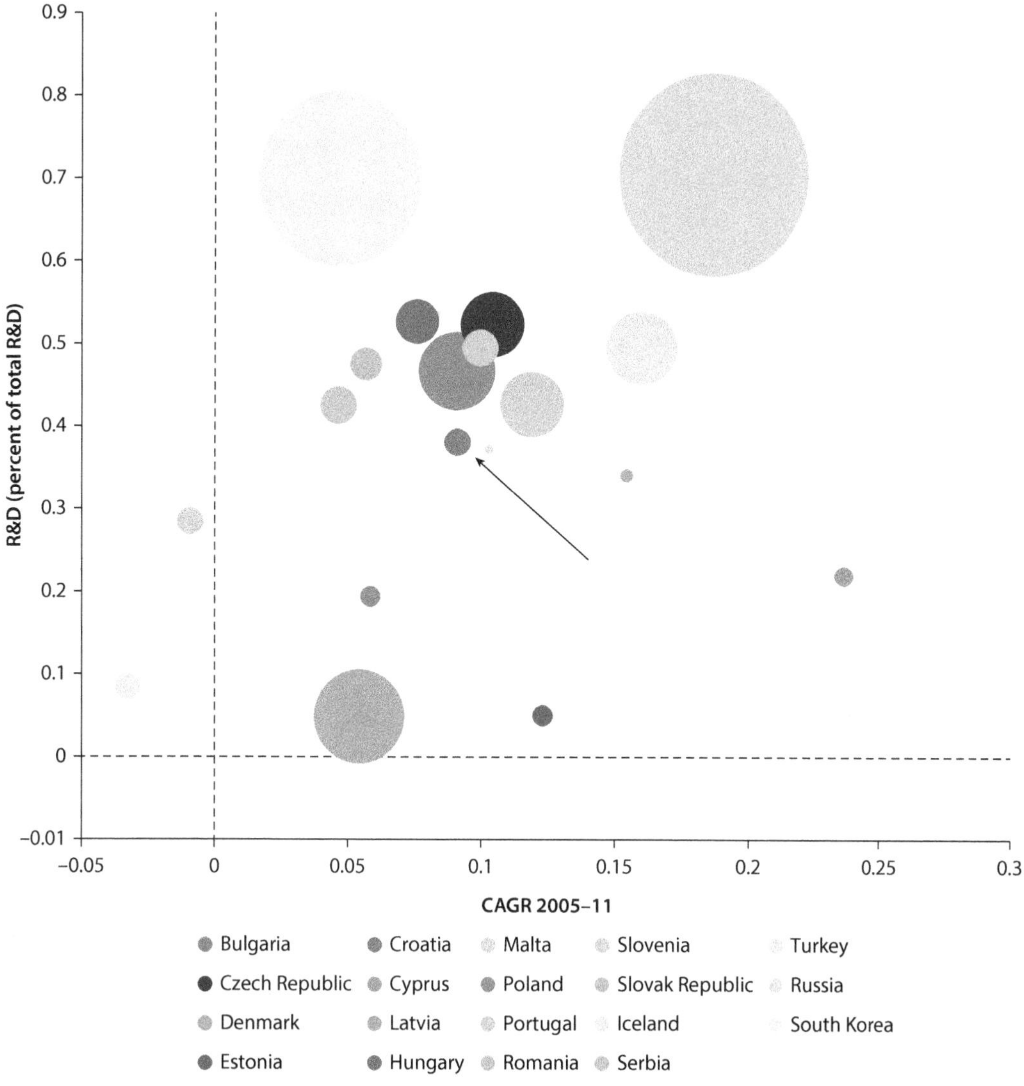

Source: Elaborations based on Eurostat data.

Oyster Production: Extremely Favorable Local Conditions[4]

Introduction

The flat oyster *Ostrea edulis*, native to Europe, has been part of the European diet for many centuries. The Romans built ponds to stock and sort oysters. However, in the past 40 years, production of *Ostrea edulis* globally has seen a drastic decline (figure A.12), due to the impact of two parasitic epizooties (*Bonamia ostreae* and *Marteilia refringens*) in the 1960s and a consequential shift to the rearing of the Portuguese cupped oyster (*Crassostrea angulata*) and then the Pacific cupped oyster (*Crassostrea gigas*) (Goulletquer 2004).

Figure A.7 Enrollment in Engineering, Manufacturing, and Construction

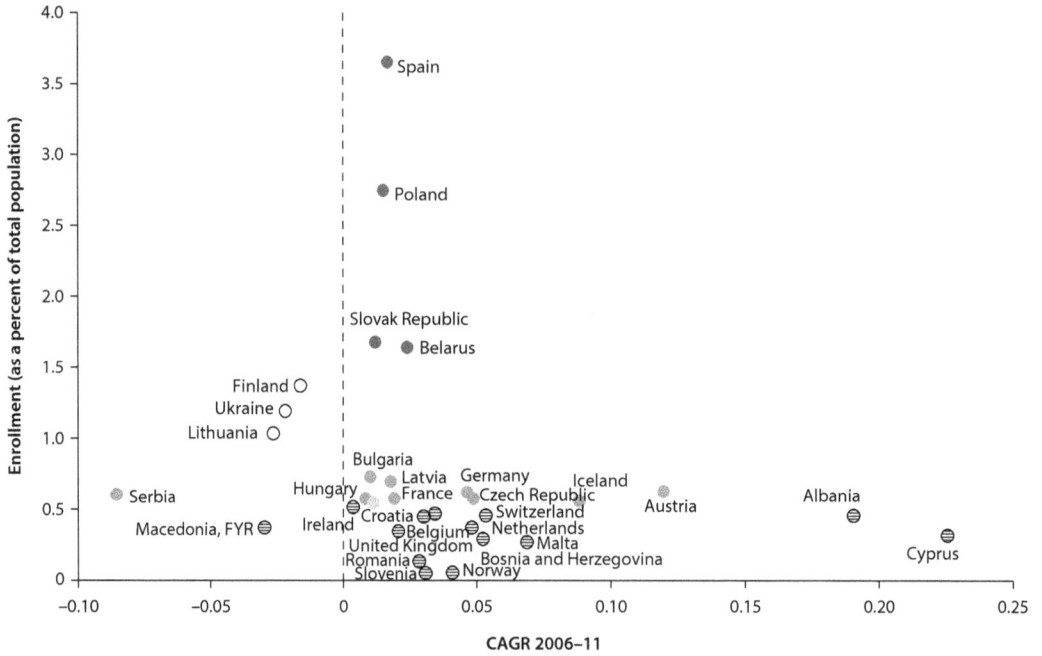

Source: Elaborations based on Eurostat data.

Figure A.8 High-Tech Exports, Croatia
€ million

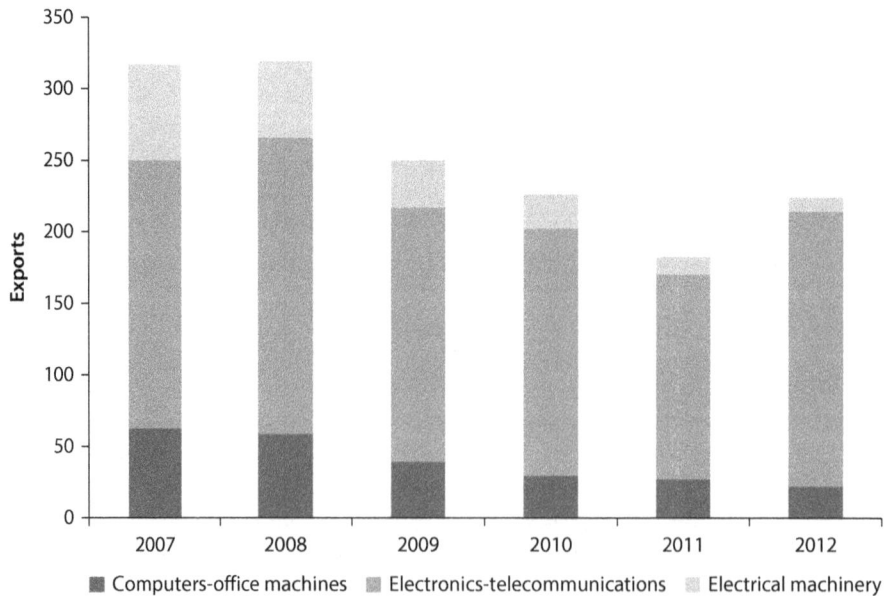

Source: Eurostat.

Figure A.9 Croatia's High-Tech Exports vs. Regional Peers'

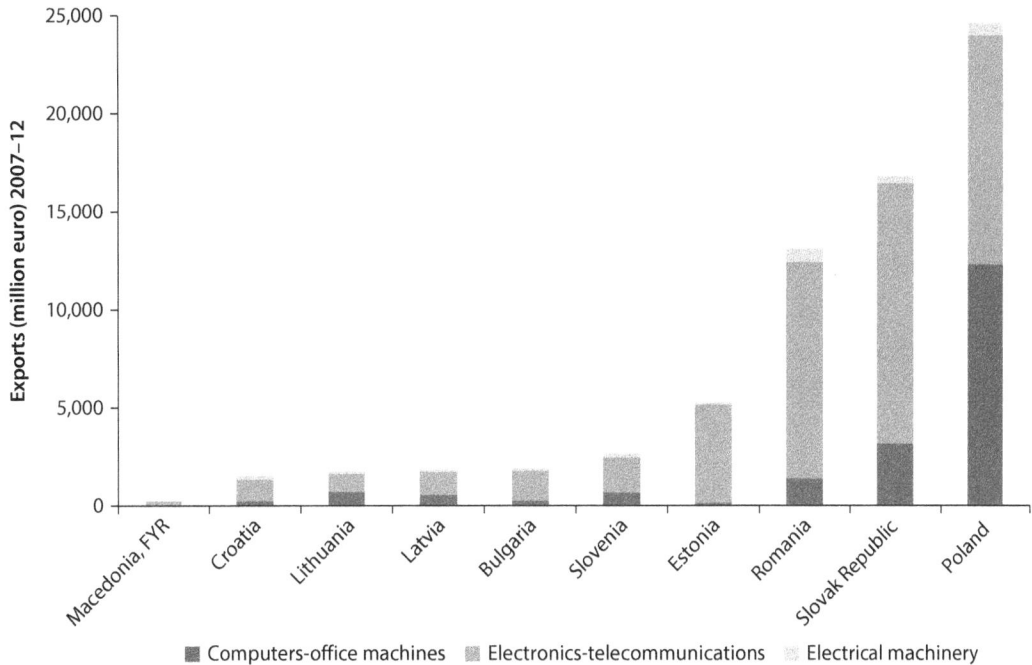

Source: Eurostat.

European flat oyster production was low throughout the decade 1993–2002; output peaked in 1996 (7,996 tons) but became more stable (6,000–7,000) in 2000, 2001, and 2002. In 2002, 67 percent of production was in Spain (4,565 tons) and 24 percent was in France (1,600 tons) (Goulletquer 2004). In 2004, Ireland and Croatia were the other countries that produced more than 200 tons. The production of European flat oyster represented less than 0.11 percent of the total global production of all farmed oyster species in 2004. Catches of wild O. *edulis* represent 10–30 percent of the total tonnage of oysters marketed in recent years (figure A.13). The bulk of world oyster production (96.2 percent) was the cultured Pacific cupped oyster (Lapègue et al. 2006).

Market and Trade Trends

Prices for all oyster species have doubled over the past three years as a result of high disease-related mortality. In spite of the price increases, demand for oysters is good, which is generating support for oyster farming projects by governments in several South and Central American countries, including Chile, Ecuador, and Republica Bolivariana de Venezuela (FAO 2013).

As supply has decreased, average prices have risen sharply (figure A.14). Although fluctuating widely, depending on size and local availability, prices have reached a record high (US$13/kg) in traditional markets in France, a key importer of oysters. The wholesale average price for flat oysters is commonly three to five

Figure A.10 Export Composition

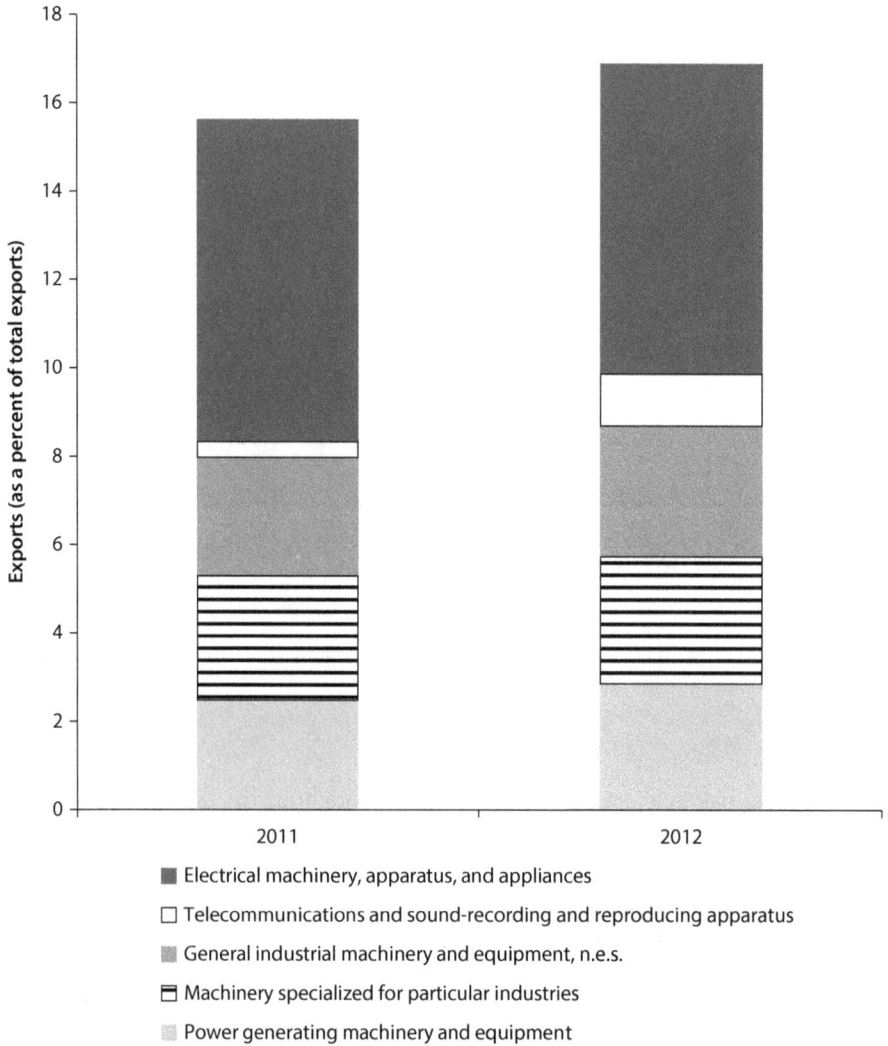

Source: Croatian Bureau of Statistics.

times higher than that for the Pacific cupped oyster. The product now therefore occupies an economic niche and is considered a luxury seafood item—an expensive delicacy for specialized consumers. The value of farmed flat oyster production was US$24.3 million in 2002 and was US$20.3 million in 2004, and so its culture remains an important industry in the few areas where it is reared (Goulletquer 2004; Lapègue et al. 2006).

Croatian Production of Flat Oysters

Shellfish farming in Croatia is still small-scale and very traditional. The country has no big shellfish farms; the few relatively large ones produce 100–200 tons

Table A.2 Significant and Emerging Exports

HS 6-digit	Product	Status	RCA 2002	RCA 2012
850423	Liquid dielectric transformers having a power handling capacity >10000kVA	Significant	28.4	48.6
730890	Structures (excld. prefabricated buildings of heading 94.06) & parts of structures (e.g., Bridges & bridge-sections, lock-gates, towers, lattice masts, roofs, roofing frame-works, doors & windows & their frames & thresholds for doors, shutters, etc.)	Significant	3.4	3.4
680610	Slag wool, rock wool & similar mineral wools (incl. intermixtures thereof), in bulk/sheets/rolls	Significant	20.4	48.7
853329	Fixed electrical resistors (excl. fixed carbon resistors, composition/film types), n.e.s. in 85.33	Significant	15.6	97.2
903289	Automatic regulating/controlling Instr. & apparatus, n.e.s. in 90.32	Significant	4.8	3.8
850300	Parts suit. for use solely/principally with the machines of 85.01/85.02	Significant	1.4	2.7
850431	Electrical transformers (excl. dielectric) having a power handling capacity not >1kVA	Significant	2.1	8.9
853650	Switches other than isolating switches & make-&-break switches, for a voltage not>1000v	Significant	1.1	2.3
850421	Liquid dielectric transformers having a power handling capacity not >650kVA	Significant	26.4	15.7
730900	Reservoirs, tanks, vats & similar containers for any material other than compressed/liquefied gas, of iron/steel, of a capacity >300 l, whether/not lined/heat-insulated but not fitted with mechanical/thermal equip.	Significant	2.7	4.5
854460	Electric conductors (excl. of 8544.11–8544.30), for a voltage >1000V	Significant	1.5	2.2
761090	Aluminum Structures (excld. prefabricated buildings of heading 94.06) & parts of structures (e.g., Bridges & bridge-sections, lock-gates, towers, lattice masts, roofs, roofing frame-works, shutters, balustrades, pillars & columns) aluminum plates	Significant	1.3	2.2
850164	AC generators (alternators), of an output >750kVA	Significant	4.5	3.6
730820	Towers & lattice masts of iron/steel	Significant	4.0	3.5
850422	Liquid dielectric transformers having a power handling capacity >650kVA but not >10000kVA	Significant	10.9	8.4
701939	Webs, mattresses, boards & similar nonwoven products of glass fibers	Significant	4.8	5.8
850152	AC motors (excl. of 8501.10 & 8501.20), multiphase, of an output >750W but not >75kW	Significant	1.5	1.0
853720	Boards, panels, consoles, desks, cabinets & other bases, equipped with 2/more apparatus of 85.35/85.36, for electric control/distribution of electricity, incld. those incorporating instruments/apparatus of Ch. 90 & numerical control apparatus, o	Significant	2.7	1.1
850790	Parts of the electric accumulators & separators therefor of 85.07	Significant	3.1	2.0
850161	AC generators (alternators), of an output not >75kVA	Significant	4.6	4.2
382450	Non-refractory mortars & concretes	Significant	6.7	2.1
853710	Boards, panels, consoles, desks, cabinets & other bases, equipped with 2/more apparatus of 85.35/85.36, for electric control/distribution of electricity, incld. those Incorporating instruments/apparatus of Ch. 90 & numerical control apparatus, o	Emerging	0.6	1.3
853321	Fixed electrical resistors (excl. fixed carbon resistors, composition/film types), for a power handling capacity not >20w	Emerging	0.4	5.7
841090	Parts (incl. regulators) of the hydraulic turbines & water wheels of 8410.11–8410.13	Emerging	0.4	3.2

table continues next page

Table A.2 Significant and Emerging Exports (continued)

HS 6-digit	Product	Status	RCA 2002	RCA 2012
850162	AC generators (alternators), of an output >75kVA but not >375kVA	Emerging	0.6	5.7
681091	Prefabricated structural components for building/civil engineering, of cement/concrete/artificial stone, whether/not reinforced	Emerging	1.0	2.6
850151	AC motors (excl. of 8501.10 & 8501.20), multiphase, of an output not >750W	Emerging	0.7	1.6
841581	Air-conditioning machines incorporating a refrigerating unit & a valve for reversal of the cooling/heat cycle (reversible heat pumps)	Emerging	0.3	1.2
841620	Furnace burners other than those for liquid fuel, incl. combination burners	Emerging	0.1	2.3
850163	AC generators (alternators), of an output >375kVA but not >750kVA	Emerging	0.3	5.7

Figure A.11 Product-Space Analysis

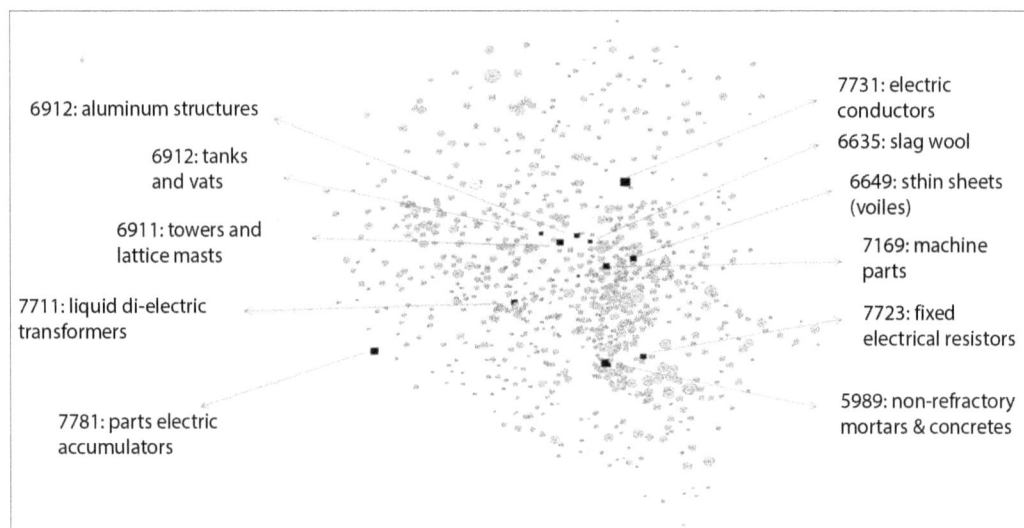

annually. There are some 120 producers in Croatian shellfish aquaculture. Production is realized through 223 shellfish farms (some farmers have more than one area of production in concession) and is limited to black mussel (*Mytilus galloprovincialis*) and European flat oyster (Piria 2012). Figure A.15 shows shellfish production in Croatia from 1990 to 2001.

Data on production vary greatly depending on source. According to the Ministry of Environmental Protection, Physical Planning, and Construction (MEPPC) (2010), total annual production is 3,000 tons of mussels and about 2 million oysters. The *Statistical Yearbook of the Republic of Croatia 2013* (Croatian Bureau of Statistics 2013) put Croatia's production of shellfish in 2012 at 1,680 tons; the Croatian Chamber of Economy put it at 2,160 tons; and the Croatian Chamber of Trades and Crafts put it at 3,150 tons.

Figure A.12 Global Aquaculture Production of Flat Oysters

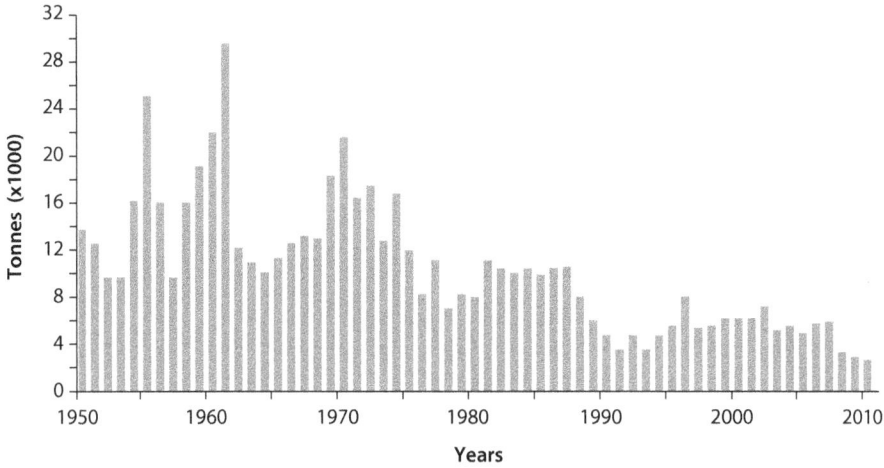

Source: FAO Fishery Statistics 2012. http://www.fao.org/statistics/en/ (reproduced with permission).

Figure A.13 Aquaculture Production of *O. edulis* in Europe by Country, 2004
Tons

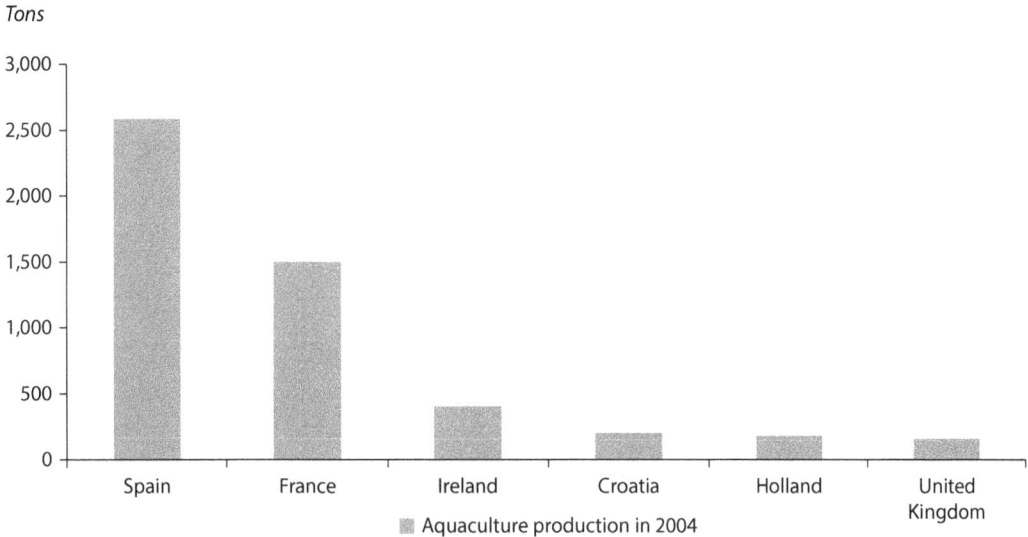

Aquaculture production in 2004

Source: FAO Fishery Statistics 2006. http://www.fao.org/statistics/en/ (reproduced with permission).

Mali Ston Bay and the Istrian Peninsula are the largest flat oyster production sites. According to MAFRD (2013), the entire yearly oyster production in Croatia in 2004 was 800,000 pieces. Only in Mali Ston Bay and Malo More region, 70 oyster farmers produce at least 2,000,000 pieces of oysters every year.

But Mali Ston Bay and Malo More seem to use only 30 percent of permitted production capacity, which points to the possibility of increasing production. Zone III (one of three zones of the production area) has many free concessions.

Figure A.14 Oyster Prices Originating in Ireland and France
Euro per kg

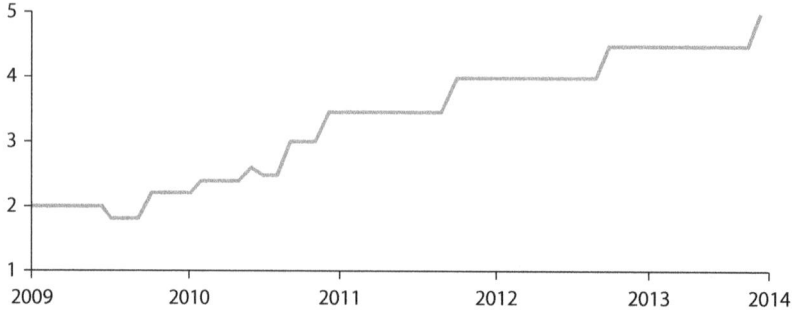

Source: FAO 2013. http://www.fao.org/docrep/019/i3473e/i3473e.pdf (reproduced with permission).

Figure A.15 Annual Production of Marine Mollusc Mariculture Industry of Croatia, 1990–2001

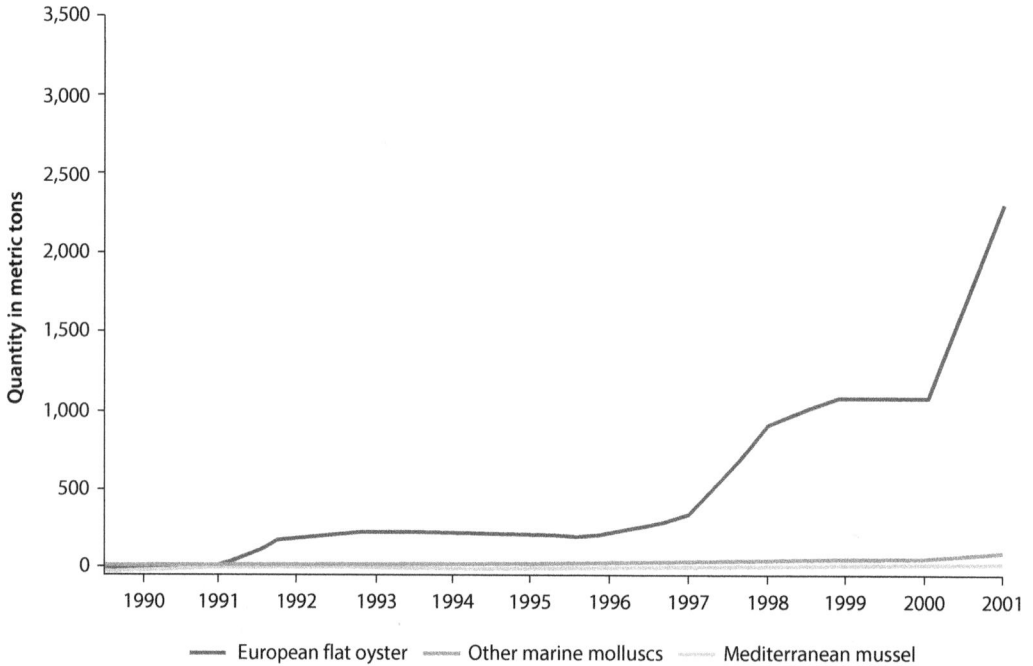

Source: FAO 2006 (Fishery Statistics) reproduced with permission.

In zones I and II, because of administrative problems (poor coordination between ministries, noncompliance with regulations), not all concessions have clear ownership. MAFRD is adjusting its maps and boundaries of concessions using data from the Croatian Hydrographic Institute, which will allow tenders to be published and these problems to be resolved.

The government has prepared documents on regulations to protect the environment, with a single location permit for the entire area. This should allow for relatively quick awarding and processing of concessions, stimulating sustainable economic development in the whole area. The main problem is that it is impossible to build facilities within the protected maritime area or to support infrastructure, such as dispatch and depuration centers. Regulations on the Protection and Preservation of the Special Reserve in the Mali Ston Bay and Malo More are also being drafted. If the MEPPC adopts proposals from representatives of shellfish producers, Dubrovnik-Neretva County, the University of Dubrovnik, and the Mariculture Business Innovation Center of the University of Dubrovnik, it will be possible to build such infrastructure on land for the shellfish industry.

Advantages of Mali Ston Bay Production Area

Croatia has extremely favorable conditions for oyster cultivation, which is no longer the case with other large producers where populations of flat oyster have been decimated because of overfishing or disease. In France, the largest producer of oysters, flat oysters were replaced by more disease-resistant Portuguese cupped oysters, but in the 1970s this species too became infected and was replaced by the Japanese oyster, which still dominates, but it has meat that is inferior to that of the European flat oyster. It is found on the eastern Adriatic coast, along the coast of Istria (Lim channel), where it presents a significant threat to the populations of indigenous flat oysters. Given that flat oysters are a prized delicacy, their preservation and even enhancement of breeding possibilities on the Adriatic coast, especially in Mali Ston Bay, where it is still the only oyster species present, offer great economic potential.

Such regional potential arises because, despite practices to limit mortality and boost flat oyster growth, diseases have hit both wild and cultured flat oyster populations all over the world. The main issue for most of the European industry is to develop a disease-tolerant strain, which requires a better understanding of the pathogenic mechanisms and the development of sustainable genetic-breeding programs. Although research programs have demonstrated the feasibility of mass selection, development of such a strain has yet to be commercialized. Mali Ston Bay area does not have that problem because two parasitic epizooties (*Bonamia ostreae* and *Marteilia refringens*), which decimated flat oyster production elsewhere in Europe, have not invaded the area, giving it a huge advantage.

Market for Flat Oysters and Marketing Venues in Croatia

Production of flat oysters does not meet demand, most of which is local. The broader (nonlocal) Croatian market has not been exploited, mainly because of too little and inconsistent production, underuse of the farming area, lack of marketing and development strategies (e.g., oyster quality and size have not been categorized), too few distribution and purification facilities, and inefficient producer associations (Jug-Dujaković and Gavrilović 2011). Most flat oysters are sold to local restaurants only.

Oyster production goes only to the domestic market. The EU market should provide opportunities for further development, but it requires modern farming technologies. Oyster cultivation is based solely on the collection of larvae naturally, as Croatia has no commercial shellfish hatcheries. These are needed to produce enough spat to ensure consistent production and create a base for expanding into new markets (MAFRD 2013).

Shellfish output stays on the domestic market to try to satisfy demand during the tourist season, but there is always a lack of supply at the end of the season. Consequently production cannot meet the basic requirements for export, which is becoming an increasing problem for the industry.

The wholesale price for flat oysters ranges from €0.40 to €0.60 each. The retail price is almost similar to the wholesale price with a larger variation (the reason for which is not readily apparent, though no doubt has something to do with supply and demand). Restaurant prices range from €1.20 to €2.60 each.

A long-term strategic objective is to achieve efficiency comparable to that of international competitors, while meeting the most stringent quality and environmental standards. Categories and brands are missing, and should be created as soon as possible to add value to what can be marketed as an exclusive product (Jug-Dujaković, Gavrilović, and Jug-Dujaković 2008; Gavrilović, Jug-Dujaković, and Skaramuca 2010a, 2010b). This requires labeling and certification of products as well as marketing material, and information on overall shellfish farming activities and their sustainability.

Certification represents a whole range of possibilities adding value to the basic product by identifying characteristics of the production process or the product itself, allowing the product to stand out from others in the market. Beyond categorization, certification requires an association of producers, especially for geographic origin and organic production. Cooperation among producers is needed to promote the association, establish minimum quality standards, and develop marketing channels within the industry and a broader marketing strategy, which builds on regional and cultural attributes.

Slavonski Kulen: Creating Value through Specializing Locally[5]

Market Analysis
Slavonski kulen, a traditional and popular gourmet pork product in Croatia, is a useful example of the potential development of a local RIS3, according to sector analysis. The kulen market has been rapidly expanding for the last decade. Industrial production more than doubled over 2004–13 (table A.3), although small traditional production declined slightly. There is a clear distinction between two types of production: Baranjski kulen and Slavonski kulen. Both have (national) protected geographical designation, and both are extending protection at the EU level. Today Baranjski kulen is a strong market leader in Croatia whose registered production and sales are almost 10 times bigger than registered sales of Slavonski kulen. One company produces 90 percent of Baranjski kulen, while 35 registered producers account for most of the Slavonski kulen production (in 2013).

The retail price for Slavonski kulen is comparable to the average retail price of medium-quality meat products in the EU. Its most important distribution channel remains direct sales (75 percent), followed by wholesale and retail (25 percent). A recent University of Zagreb survey confirmed that consumers

Table A.3 Kulen Production

All types of kulen production	2004		2013	
	kg	%	kg	%
1. Industrial production	129,277	79.1	300,000	90.4
2. Small traditional production	33,378	20.9	32,000	9.6
Total	162,655	100	332,000	100

Table A.4 Percentage of Participants Agreeing with Each Statement

Slavonski kulen is a:	Total (n = 1,000)	Slavonia (n = 199)	Central (n = 104)	Nord, North West	Zagreb (n = 246)	Istria, Primorje	Dalmatia (n = 193)
Traditional Croatian product	97	99	96	97	99	92	96
Gourmet product for enjoyment	87	92	82	92	87	86	84
Product for special occasions	79	83	79	86	74	75	80

tend to perceive Slavonski kulen as a traditional and popular gourmet pork product, often used for special occasions (table A.4).

Although consumer demand for Slavonski kulen remains high, the supply is small, and there is much room for growth. According to a recent survey of producers (2014), weak supply derives from multiple causes: the high price associated with the value-added tax (25 percent), which undermines the product's competitiveness; a shortage of high-quality raw materials (pig production in Croatia fell sharply after 2010, when the government ended support); strict national veterinary requirements; and limits on the number of pigs that kulen producers can have on their farms.

Croatia has a solid comparative advantage in Slavonski kulen production. It derives from Slavonia's solid agricultural base in animal feed production and from a large reservoir of positive perceptions (see table A.4), including an extraordinarily good image and reputation as a traditional product. It also enjoys stable and growing demand domestically, including tourist consumption, and it also has export potential, by which it could fill a niche in premium markets. Its protected geographical designation potentially opens new markets in Austria, Germany, and Switzerland, where it may appeal to sophisticated consumers.

Tourists in Croatia can test-market and promote kulen because producers can learn foreign consumers' preferences quite inexpensively. They can also help cultivate its reputation, including brand identification. Assuming only 5 percent of tourists consumed 50 grams of Slavonski kulen while visiting Croatia, this would translate into a potential demand for 150,000 kg of kulen a year, or half the sales in the country today.

Kulen was the first of 15 products to receive protection via protected geographical in Croatia. Six of these also have protected designation of origin status.[6,7] Some of these are set to receive EU protection as well, which entails legal obligations on product and brand names, duration of the production process, allowable weight of fattened pigs for the final product, age limits for slaughtered pigs, product fat and water content, meat granulation, and geographic areas of production.

Slavonski Kulen production has to overcome five challenges: weak veterinary and safety regulations, inadequate access to capital, lack of economies of scale, patchy horizontal and vertical cooperation, and poor marketing and promotion. The first imposes the tightest constraints. The Croatian government has already harmonized its laws and regulations with EU standards, but Slavonski kulen producers complain that they are in practice stricter than those imposed in the "old" EU member states. They relate to quality schemes, hygiene, and animal-feed controls. Second, commercial interest rates are two to three times higher in Croatia than in the old EU member states, and this impedes kulen producers' access to capital. Local banks rarely accept agricultural land as collateral; where they do, these banks typically require much larger collateral than the loan's value. This discourages any expansion, which is why producers labor under the third challenge—failure to exploit economies of scale.

Horizontal cooperation and, to a lesser extent, vertical cooperation are crucial to transforming kulen producers' small-scale business activities into more ambitious enterprises. Many farmers, however, remain averse to cooperation because they had negative experience with cooperatives during an earlier era. Only a few examples of horizontal cooperation appear, and this remains the most important type of cooperation if kulen producers hope eventually to exploit economies of scale. This requires innovation and the adoption of new technologies. Thus far, technological innovation has included a breeding program and protected designation of origin, which have both enjoyed a measure of success.

Better marketing and promotion—the fifth challenge—requires greater commitment and effort. Specialized fairs, exhibitions, and quality contests, for example, are particularly important. Slavonski kulen producers may also increase sales if they improve the packaging.

Policies to Support Slavonski Kulen

The government has offered many types of support for Slavonski kulen production. A pioneer project, Export Marketing of Domestic SK [Slavonski kulen] (2001–03), cofinanced by the Ministry of Agriculture and Forestry, is one. This project heightened awareness of the need to standardize the product and cultivate its reputation.

MSES cofinanced a technological research project called Slavonian Domestic Kulen from the HITRA-TEST[8] program. It followed up activities initiated by the Agricultural Faculty of Zagreb. The main objective was to modernize the traditional technologies in producing Slavonski kulen. The most suitable genotype of pigs was selected, the current quality-control system for fresh meat was determined, production methods in traditional technology were stipulated, and the quality of the product was established. The project ended in 2005.

Government support has remained particularly diverse since 2000. Most subsidies supported production. The range of covered products and the budget increased hugely. However, this diversity in subsidies necessarily resulted in market distortions. Some productive activities became stagnant, while subsidies kept other unproductive ones alive.

This government support rendered some farmers dependent on subsidies by the mid-2000s. But any signal that the subsidies might be reduced or limited led to strong protests—after which the government increased subsidies. The value added in agriculture generally did not reach the level that policies had envisaged: indeed, production stagnated, and self-sufficiency worsened. Therefore, the government started aiming to increase livestock production in 2005, focusing on milk, pigs, beef and veal, perennial crops, and Slavonski kulen, with its own operational program.

Biotechnology and Pharmaceuticals: High Risks and Uncertain Returns[9]

Agriculture, food, and beverages are among the largest and most successful branches of the Croatian economy (see chapter 3). They represent a large share of GDP, the labor force, and exports and still offer great expansion potential. Biotechnology and pharmaceuticals, too, are a "hotspot" with key technology important for competitiveness (table A.5), primarily in health care, food processing, and agribusiness. While visiting Croatia in 2013, EU Commissioner Johannes Hahn stated, "There appears to be big potential for developing the bio-energy, bio-technology and bio-health sector."[10]

Yet Croatia lacks a strategy for developing biotechnology and pharmaceuticals. The government is, though, crafting, updating, or aiming to adopt several policies that will directly or indirectly boost these two sectors, including the following:

- A new Strategy for Education, Science, and Technology (under public discussion).
- A Croatian Innovation Strategy.
- The RIS3 (in preparation).
- An Action Plan for the Development of Agriculture in Croatia (2011–16).

Table A.5 Performance in Research, Innovation, and Competitiveness in Croatia

	Investment and Input		Performance/economic output	
Research	R&D intensity		Excellence in S&T	
	2011: 0.75%	(EU: 2.03%; US: 2.75%)	2010: 12.25	(EU: 47.86; US: 56.68)
	2000–2011: −2.72%	(EU: +0.8%; US: +0.2%)	2005–2010: + 2.31%	(EU: +3.09%; US: +0.53)
Innovation and Structural change	Index of economic impact of innovation		Knowledge-intensity of the economy	
	2010–2011: 0.353	(EU: 0.612)	2010: n.a	(EU: 48.75; US: 56.25)
			2000–2010: n.a.	(EU: +0.93%; US: +0.5%)
Competitiveness	Hot-spots in key technologies		HT + MT contribution to the trade balance	
	Health care sector; Food processing and		2011: 2.98%	(EU: 4.2%; US: 1.93%)
	agribusiness; Energy technology; Electronics and		2000–2011: +133.23%	(EU: +4.99%; US: −10.75%)
	Advanced materials and Digital techniques			

Source: European Commission 2013.

Research Output in Health Care, Biopharmaceuticals, and Agriculture: Croatia and Three European Leaders

Historically, Croatia has had good research output and internationally recognized scientists in medicine, life sciences, and chemistry, including two Nobel laureates. Vladimir Prelog, a Croatian-Swiss organic chemist, who received the 1975 Nobel Prize in Chemistry, lived and worked in Prague, Zagreb, and Zürich during his career. Leopold Ružička, a Croatian scientist and winner of the 1939 Nobel Prize in Chemistry, worked most of his life in Switzerland.

In 1980, Croatian chemists at the pharmaceutical company Pliva, d.d. Zagreb, discovered the antibiotic azithromycin. It was patented in 1981. Pliva and Pfizer signed a licensing agreement in 1986, whereby Pfizer received exclusive sales rights in Western Europe and the United States, while Pliva began marketing the drug in Central and Eastern Europe under the brand name Sumamed in 1988.

The selection of biotechnology-related sectors, such as health and agribusiness, as hotspots is supported by analysis of the scientific research output of Croatian R&D. The number of scientific publications by Croatian scientists in fields important for developing both biotechnology and pharmaceuticals exceeded that in other areas in 2011–12 (figure A.16).

The SCImago database uses categories that are internationally recognized, drawing on the Field of Science and Technology classification to enable benchmarking between countries. The categories are total published documents, citable documents, citations, self-citations, citations per document, and an *h*-index (table A.6). The first two categories measure the quantity, the next four measure the quality, of the output. The *h*-index attempts to measure both the productivity and the impact of the published work of a scientist or scholar; it is based on the set of the scientist's most cited papers and the number of citations

Figure A.16 Areas of Scientific Output in Croatia, According to Number of Scientific Articles, 2011–12

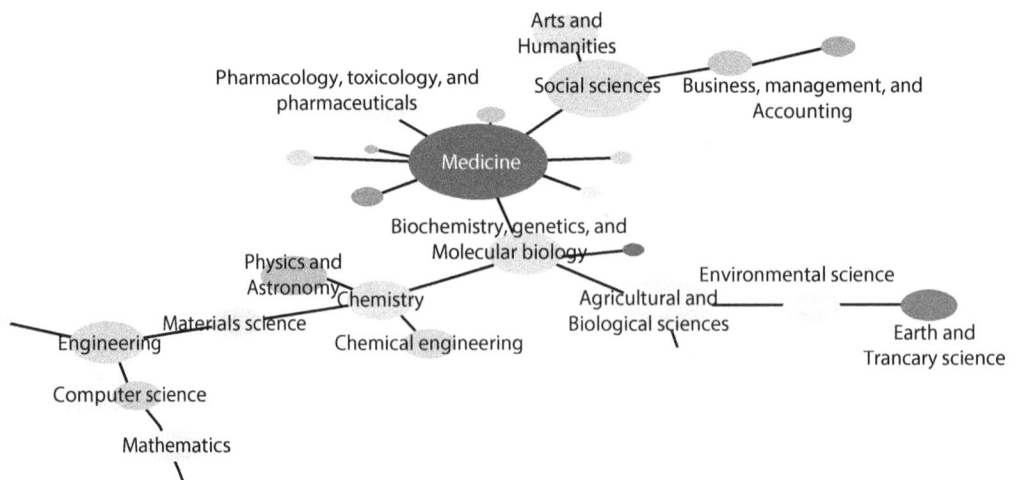

Source: SCImago 2014.

Table A.6 Croatia's Ranking among 23 Eastern European Countries in Areas of Science

Area of science	Published documents	Croatia's rank in Southeast Europe (based on number of docs.)	Citable documents	Citations	Self-citations	Citations per document	h-Index
Biochemistry, genetics, and molecular biology	4,408	10	4,248	51,095	8,893	13.25	82
Chemistry	4,263	11	4,232	36,859	10,319	9.34	58
Immunology & microbiology	954	9	893	13,570	1,933	16.46	50
Medicine	15,046	5	14,182	77,343	14,975	5.94	86
Pharmacology,toxicology, pharmaceuticals	2,055	5	2,008	11,795	3,112	6.56	43
Veterinary medicine	807	5	806	2,763	660	4.16	21
Agriculture and biosciences	6,214	6	6,112	31,238	8,847	6.08	55

Source: SCImago database.

received in other publications. The h-index can also be applied to the productivity and impact of a group of scientists, such as a department, university, or country, as well as a scholarly journal. (It was put forward by Jorge E. Hirsch and is thus sometimes called the Hirsch index or Hirsch number.)

Croatia ranks quite strongly in medicine, according to the SCImago database of May 2014, among 233 countries worldwide: at 43 on the number of publications; at 44 on the number of citations; and at 46 on the h-index. In biochemistry, genetics, and molecular biology, of 218 countries on the number of published articles it ranks at 50; on the number of citations, it ranks at 44; and on the h-index, it ranks at 45. In pharmacology, toxicology, and pharmaceuticals, of 202 countries on the number of published articles it ranks at 42; on the number of citations, it ranks at 44; and on the h-index, it ranks at 43. In agriculture and biological sciences, of 230 countries on the number of published articles it ranks at 43; on the number of citations it ranks at 43; and on the h-index, it ranks at 53. May 2014's SCImago database was also used to compare Croatia with 22 other Eastern European countries[11] for 1996–2012 in seven areas of science (table A.6).

On the number of published documents Croatia ranks best in medicine; pharmacology, toxicology, and pharmaceuticals; and veterinary medicine. Qualitatively, medicine appears the most productive and the highest quality, on the basis of the number of citations and h-index. Although biochemistry, genetics, and molecular biology rank at 10 on the basis of the number of papers, these areas have the second-highest number of citations and a high h-index (82). Chemistry is in third place on the basis of the number of citations and h-index.

Thus the two areas of medicine and of biochemistry, genetics, and molecular biology show the highest productivity in Croatia. With chemistry, these two areas generate the best quality scientific data (viz. the number of citations and h-index). Veterinary medicine ranks high in the number of papers but not in the quality of publications.

Smart Specialization in Croatia • http://dx.doi.org/10.1596/978-1-4648-0458-8

Medicine

On the basis of the number of citable documents, Croatian scientists published significantly fewer articles in medicine than scientists in Switzerland and Austria, both being countries (along with Slovenia) that ranked highly in this area.[12] The first two countries showed a steady increase in the number of citable scientific documents over 2002–12, though with Switzerland producing more (figure A.17). Croatia, with a population of 4.2 million and around 7,000 researchers, has since 2002 published more scientific articles than Slovenia— perhaps surprising given that it invests only 0.75 percent of GDP in R&D versus Slovenia's 2.47 percent. As R&D in medicine is very expensive, with more investment, Croatia should have strong R&D potential in medicine, which could well be an excellent basis for technology transfer and developing biotechnology and biopharmaceuticals. Yet despite the fewer citable medicine research articles in Slovenia, the quality of articles as shown by the h-index in Croatia and that of Slovenia are very similar (figure A.18).

Biochemistry, Genetics, and Molecular Biology

As with medicine, Croatia (like Slovenia) is significantly behind Austria and Switzerland in these very important areas for developing biotechnology and pharmaceuticals. But in biochemistry, genetics, and molecular biology, Croatia and Slovenia have an almost identical number of papers published (figure A.19), unlike in medicine, and even though Slovenia has more full-time equivalent researchers (8,774 in all sectors, 2011) than Croatia (6,847 in all sectors, 2011)

Figure A.17 Number of Citable Documents in Medicine, Austria, Switzerland, Croatia, and Slovenia, 1996–2012

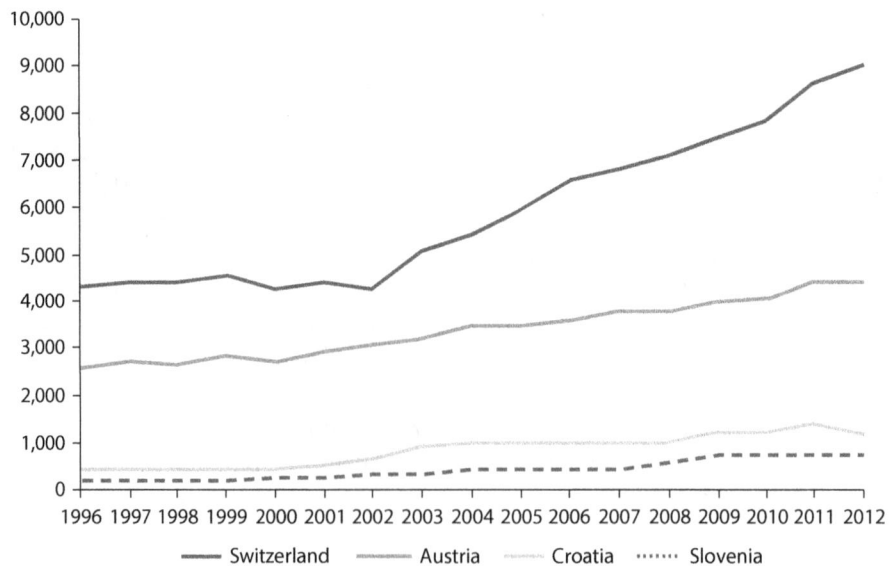

Source: SCImago Journal & Country Rank 2014.

Figure A.18 Quality of Scientific Output in Medicine as Measured by the *h*-index, Austria, Switzerland, Croatia, and Slovenia, 1996–2012

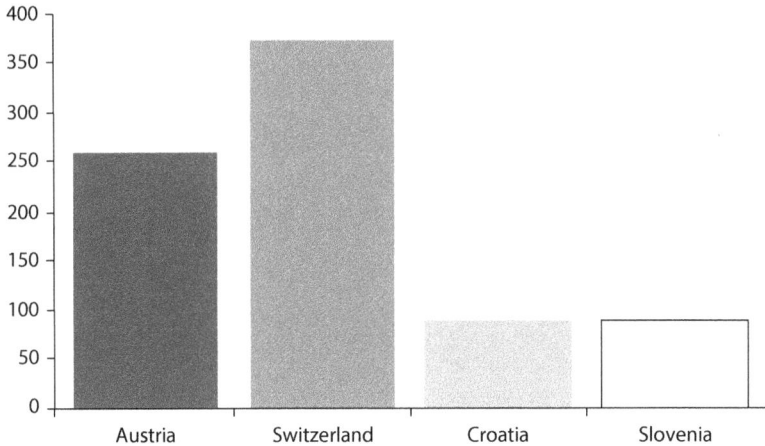

Source: SCImago Journal & Country Rank 2014.

in these scientific disciplines. These areas all involve the need for modern infrastructure and investment in laboratory consumables. Slovenia invests close to four times as much in R&D as a share of the GDP than Croatia. Slovenia has been in the EU for longer than Croatia and, since 2004, has received substantial funding for these two elements. These factors may explain the same number of citable documents and similar quality of publications as measured by the *h*-index (figure A.20).

Pharmacology, Toxicology, and Pharmaceuticals
Scientific productivity in these areas is directly linked to the potential for developing biotechnology and pharmaceuticals. All four countries have shown a trend increase in publication output in these areas since 2000, especially Switzerland, creating a wide gap with the other countries. Croatia publishes more than Slovenia, but much less than Austria and Switzerland (figure A.21), but its quality is very similar to Slovenia's as measured by the *h*-index (figure A.22). The difference in number of publications among Austria, Croatia, and Slovenia is much less than that in biochemistry, genetics, and molecular biology.

Agriculture and Biological Science
Croatia ranks very high in these fields, notably in food science, aquatic science, and agronomy and crops science. Once again Switzerland and Austria publish more than Croatia (in turn more than Slovenia; figure A.23). On the *h*-index, Switzerland as before leads the field (figure A.24).

Among the countries of Eastern Europe, Croatia has very good potential in agronomy and crops science, aquatic science, food science, and horticulture.

**Figure A.19 Number of Citable Documents in Biochemistry, Genetics, and Molecular
Biology, Austria, Switzerland, Croatia, and Slovenia, 1996–2012**

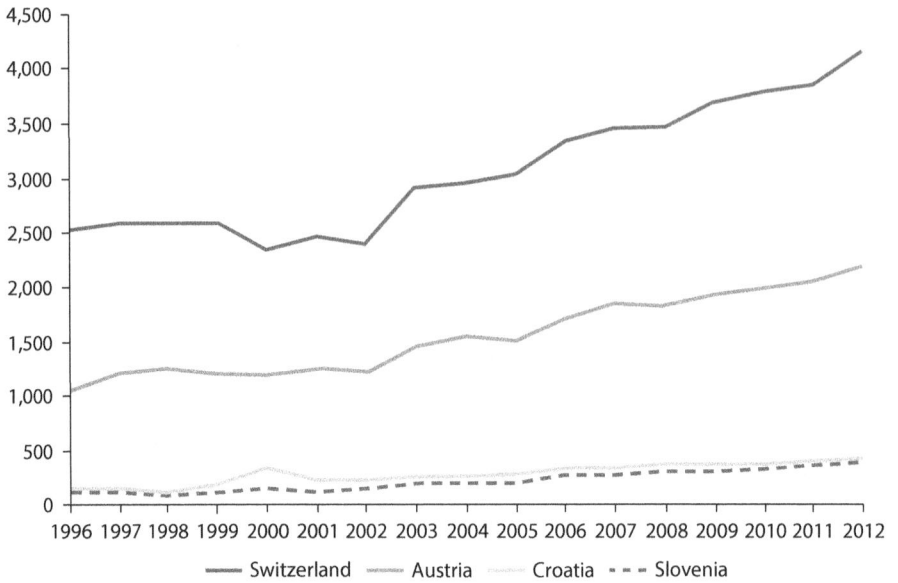

Source: SCImago Journal & Country Rank 2014.

**Figure A.20 Quality of Scientific Output in Biochemistry, Genetics, and
Molecular Biology as Measured by the *h*-Index, Austria, Switzerland, Croatia,
and Slovenia, 1996–2012**

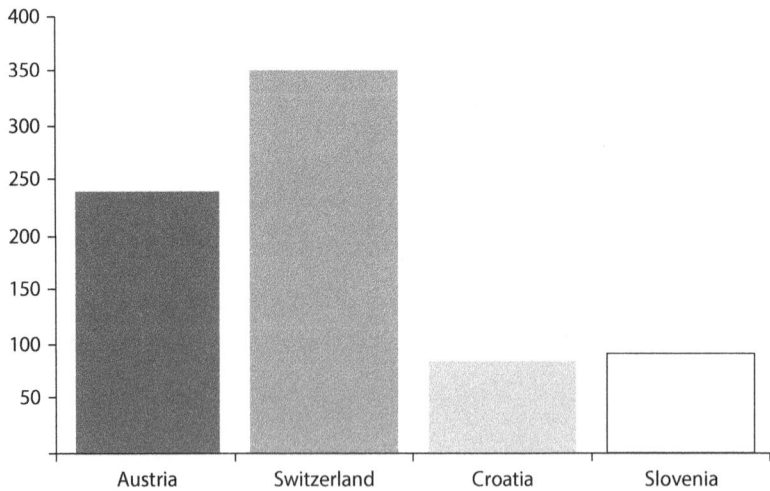

Source: SCImago Journal & Country Rank 2014.

Figure A.21 Number of Citable Documents in Pharmacology, Toxicology, and Pharmaceuticals, Austria, Switzerland, Croatia, and Slovenia, 1996–2012

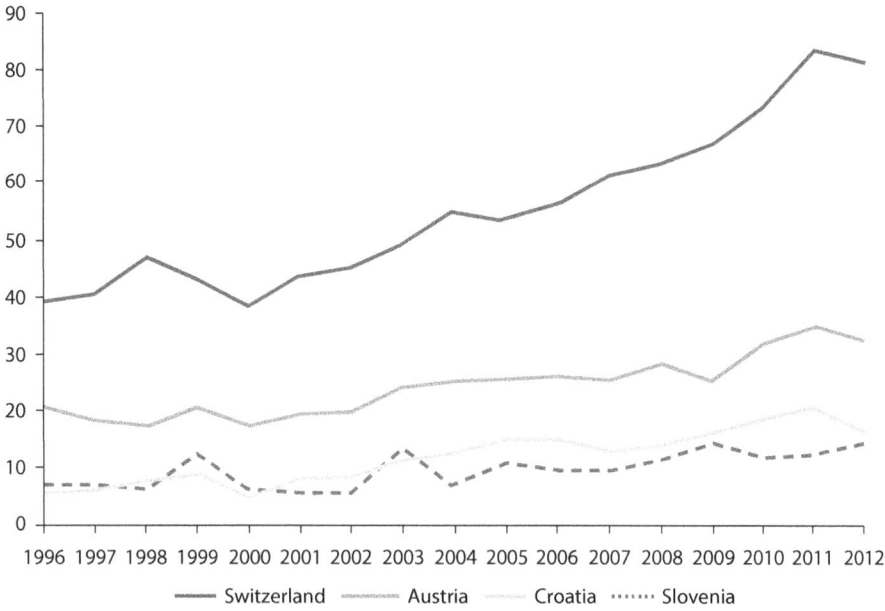

Source: SCImago Journal & Country Rank.

Figure A.22 Quality of Scientific Output in Pharmacology, Toxicology, and Pharmaceuticals as Measured by the *h*-Index, Austria, Switzerland, Croatia, and Slovenia, 1996–2012

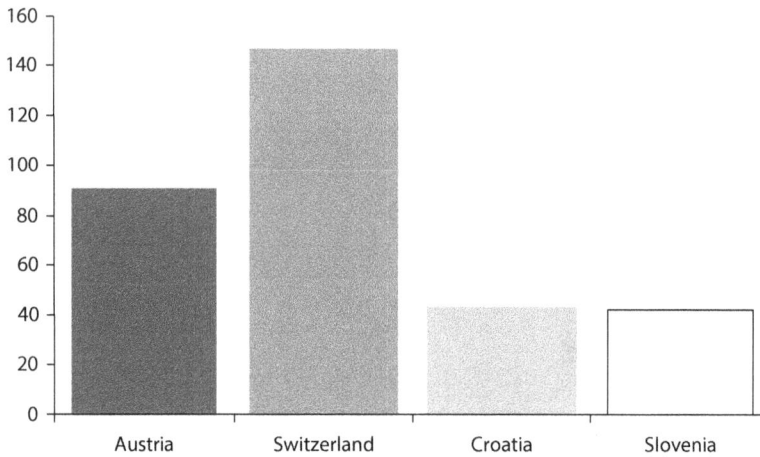

Source: SCImago Journal & Country Rank.

Figure A.23 Number of Citable Documents in Agriculture and Biological Sciences, Austria, Switzerland, Croatia, and Slovenia, 1996–2012

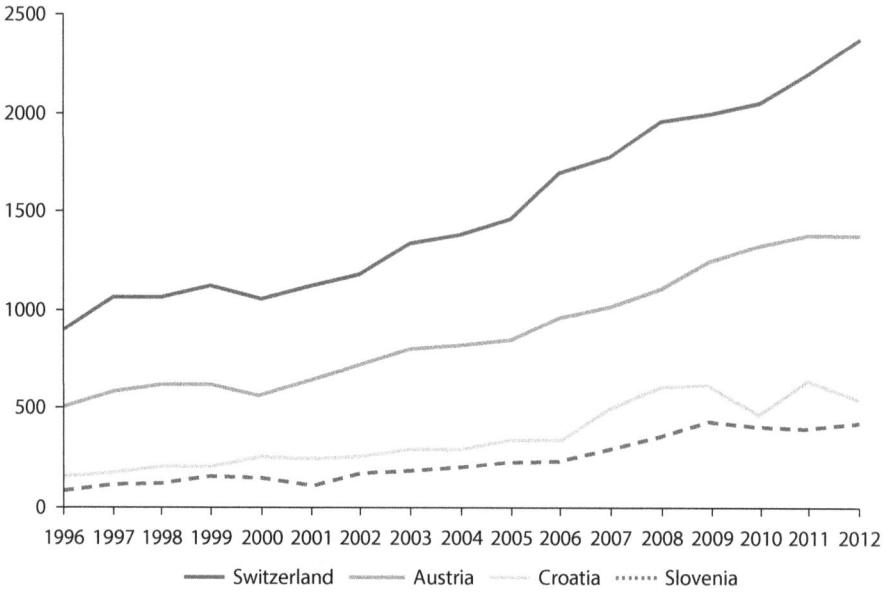

Source: SCImago Journal & Country Rank.

Figure A.24 Quality of Scientific Output in Agriculture and Biological Sciences as Measured by the *h*-Index, Austria, Switzerland, Croatia, and Slovenia, 1996–2012

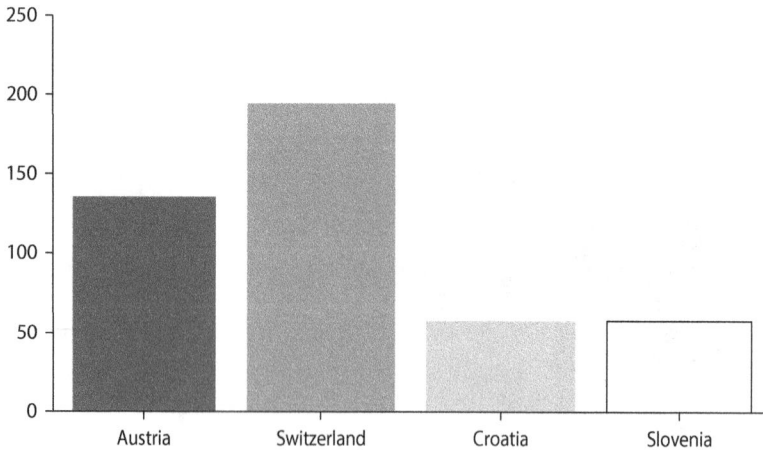

Source: SCImago Journal & Country Rank.

This should be considered in the future planning of the development of blue[13] and green biotechnology. Yet even in these four areas, the h-indexes for Austria and Switzerland are still much higher than in Croatia (data not shown). Thus despite its potential, Croatia still needs to improve quality in this area substantially, and may wish to consider steps along the following lines:

- The government should develop a clear strategy and action plan for developing this sector. A Croatian Biotechnology and Pharmaceuticals Initiative with clear vision and appropriate measures should be developed.
- A biotechnology and pharmaceutical cluster including both the research and private sectors should be set up. A small management company with experience in this sector should coordinate future cluster activities operationally.
- MSES should open a wide swathe of postdoctoral positions in biotechnology fields and attract top scientists from Croatia and abroad.
- The Croatian biotechnology and pharmaceutical sector should be promoted internationally (EU, United States, bio conferences, through venture capital, and so on).
- Strategic international partners and investors should be attracted, by means including special incentives.
- Top-level researchers should be attracted to work in Croatia (special packages should be provided).
- There should be high-quality PhD programs in English in the areas of science that are important for biotechnology development.

Notes

1. The authors of this assessment are Gabriela Elizondo Azuela and Asad Ali Ahmed.
2. In this section we focus on renewable energy (all types), energy efficiency and smart grids, carbon capture and storage, and electric vehicles (largely avoiding discussion of fossil fuels, fuel economy, and nuclear power).
3. Some of these products are ranked among the world's most complex exports according to the Product Complexity Index computed by the Observatory of Economic Complexity. For example, specialized industry machinery and parts are ranked at 1; switchboards, relays, and fuses are ranked at 127; and electrical transformers are ranked at 363.
4. Inputs for this section have been provided by Jurica Jug-Dujaković and Ana Gavrilović.
5. Inputs for this section have been provided by Miroslav Božić.
6. Such a designation means that only products originating in a specific region are allowed to be identified as such.
7. Neretva mandarins, Dalmatia Maraska sour cherry, Ogulin sour cabbage, Cres extra virgin olive oil, Istarski prsut, and Varazdin cabbage.
8. http://erawatch.jrc.ec.europa.eu/erawatch/opencms/system/modules/com.everis .erawatch.template/pages/exportTypesToHtml.jsp?contentid=488e0a55-9095-11e0 -a33b-3b1a37daf5b5&country=Croatia&option=PDF.
9. Inputs for this section have been provided by Danica Ramljak.

10. http://europa.eu/rapid/press-release_SPEECH-13-117_en.htm.

11. They are Albania, Armenia, Azerbaijan, Belarus, Bosnia and Herzegovina, Bulgaria, the Czech Republic, Estonia, Lithuania, Latvia, Georgia, Hungary, Macedonia, Moldova, Montenegro, Poland, Romania, Russia, Serbia, Slovenia, the Slovak Republic, and Ukraine.

12. Switzerland: population 8 million; invests 2.87 percent of GDP in R&D; Austria: population 8.5 million, invests 2.75 of GDP in R&D; Slovenia: population 2 million, invests 2.7 percent of GDP in R&D.

13. This is a term referring to the application of molecular biological methods to marine and freshwater life.

Bibliography

Bloomberg New Energy Finance. 2013. "Global Trends in Clean Energy Investment." http://about.bnef.com.

Croatian Bureau of Statistics. 2013. *Statistical Yearbook of the Republic of Croatia 2013*. Zagreb: Croatian Bureau of Statistics.

European Commission. 2012. *Guide to Research and Innovation Strategies for Smart Specialisations*. European Commission.

———. 2013. *Research and Innovation Performance in EU Member States and Associated Countries*. European Commission.

European Patent Office database. 2014. http://ep.espacenet.com.

Eurostat database. 2014. http://ec.europa.eu/eurostat/data/database.

FAO (Food and Agriculture Organization of the United Nations). 2006. *Fishery Statistics*. http://www.fao.org/statistics/en/.

———. 2012. *Fishery Statistics*. http://www.fao.org/statistics/en/.

———. 2013. *Food Outlook: Biannual Report on Global Food Markets*. Trade and Markets Division Rome, 133.

Gavrilović, A., J. Jug-Dujaković, and B. Skaramuca. 2010a. "Organic Production in Aquaculture." The First International Fishery Convention "BH-FISH 2010" Center for Fisheries "Neretva " Konjic, Bosnia and Herzegovina, June 23–24, 105–12.

———. 2010b. "Quality Labels, Certification Schemes and Market Protection of the Aquaculture Products." [Oznake kvalitete, certifikacijske sheme i zaštita proizvoda slatkovodne akvakulture na tržištu. Zbornik sažetaka Četvrtog međunarodnog savjetovanja o slatkovodnom ribarstvu Hrvatske: Hrvatsko ribarstvo kako i kuda dalje? Ribarstvo i zaštita zdravlja riba.] Vukovar 14 i 15.4. 2010. Hrvatska gospodarska komora—Sektor za poljoprivredu, poljoprivredu, prehrambenu industriju i šumarstvo, Hrvatska gospodarska komora—Županijska komora Vukovar (ur.).]

Goulletquer, P. 2004. "Cultured Aquatic Species Information Programme: Ostrea edulis." FAO Fisheries and Aquaculture Department, Rome. http://www.fao.org/fishery/culturedspecies/Ostrea_edulis/en.

IEA (International Energy Agency). 2013. *Tracking Clean Energy Progress* 2013. OECD/IEA, Paris.

Jug-Dujaković, M., A. Gavrilović, and J. Jug-Dujaković. 2008. "Possible Means of Protection and Identification of Mali Ston Oyster in the Market." *Naše more* 55 (5–6): 262–68.

Jug-Dujaković, J., and A. Gavrilović. 2011. "The Use of Advantages of the Aquaculture Production in the Modern Marketing." The Fifth Convention on the Freshwater Aquaculture, Croatian Chamber of Commerce, Vukovar, Croatia, April 7–9.

Lapègue, S., A. Beaumont, P. Boudry, and P. Goulletquer. 2006. "European Flat Oyster: Ostrea Edulis." GENINPACT: Evaluation of Genetic Impact of Aquaculture Activities on Native Population. A European network, WP1 Workshop "Genetics of Domestication, Breeding and Enhancement of Performance of Fish and Shellfish," GENINPACT Final Scientific Report, pp. 70–75, http://archimer.ifremer.fr/doc/00000 /3321.

MAFRD (Ministry of Agriculture, Fisheries, and Rural Development). 2013. "IPARD Programme 2007–2013." Agriculture and Rural Development Plan Directorate, 429.

MEPPC (Ministry of Environmental Protection, Physical Planning, and Construction). 2010. *Fifth National Communication of the Republic of Croatia under the United Nations Framework Convention on Climate Change.*

Piria, M. 2012. "National Aquaculture Sector Overview: Croatia." National Aquaculture Sector Overview Fact Sheets, FAO, Fisheries and Aquaculture Department, Rome. http://www.fao.org/fishery/countrysector/naso_croatia/en.

SCImago Journal & Country Rank database (accessed May 2014), http://www.scimagojr .com.

World Bank. 2014. *Building Competitive Green Industries: The Climate and Clean Technology Opportunity for Developing Countries.* InfoDev. Washington, DC: World Bank.

green
press
INITIATIVE

www.ingramcontent.com/pod-product-compliance
Lightning Source LLC
Chambersburg PA
CBHW082356270326
41935CB00013B/1643